FIELDWORK INVESTIGATIONS
A Self Study Guide

Sue Warn and David Holmes

Hodder & Stoughton
www.hodderheadline.co.uk

Orders: please contact Bookpoint Ltd, 130 Milton Park, Abingdon, Oxon OX14 4SB. Telephone: (44) 01235 827720. Fax: (44) 01235 400454. Lines are open from 9.00 – 6.00, Monday to Saturday, with a 24 hour message answering service. You can also order through our website www.hodderheadline.co.uk.

British Library Cataloguing in Publication Data
A catalogue record for this title is available from the British Library

ISBN 0 340 679 697

First Published 2003
Impression number   10 9 8 7 6 5 4 3 2 1
Year   2007 2006 2005 2004 2003
Copyright © 2003 David Holmes and Sue Warn.

Typeset by Pantek Arts.
Printed by J.W. Arrowsmith Ltd, Bristol for Hodder & Stoughton Educational, a division of Hodder Headline Plc, 338 Euston Road, London, NW1 3BH.

CONTENTS

INTRODUCTION 2

1. SKILLS 7
   Sampling 7
   Questionnaires 9
   Mapping 17
   Graphical skills 18
   Using photographs 21

2. PHYSICAL ENVIRONMENTS 27
   RIVERS AND VALLEYS 27
   River channel variables: basic survey techniques 28
   Site selection 35
   River water quality 41
   River and catchment management 47
   River valleys and slope surveys 49
   Hydrological projects 52
   COASTS 55
   Erosional studies 56
   Depositional studies 61
   Coastal management surveys 65
   ICE DETECTIVE 72
   Ice projects 73
   Distribution and orientation of features 73
   Morphological mapping 76
   Sediment analysis 78
   Weathering studies 83
   Mountain projects 84
   Activity and amenity surveys 84
   WEATHER 86
   Weather projects 87
   Weather station studies 91
   Microclimate studies 94
   ECOSYSTEMS 102
   Investigating soils 103
   Investigating vegetation 112
   Common British ecosystems 115
   Assessing the human impact on ecosystems 126

3.	HUMAN ENVIRONMENTS	132
	RURAL INVESTIGATIONS	132
	Rural land use surveys	133
	Farm surveys	136
	Investigating rural settlements	140
	URBAN ENVIRONMENTS	147
	The CBD	148
	Shopping studies	153
	Investigating residential land use	155
	Urban catchment area or hinterland surveys	164
	Population and people surveys	167
	POLLUTION OF THE ENVIRONMENT	180
	Investigating air pollution	180
	Investigating noise pollution	187
	Land based pollution	188
	TRANSPORT INVESTIGATIONS	192
	Traffic flow surveys	192
	Public transport surveys	197
	Route quality surveys	200
	Parking surveys	201
	Investigating the impact of transport terminals	201
	SPORT LEISURE, RECREATION AND TOURISM	203
	Basic techniques	204
	Example projects	211

INTRODUCTION

The purpose of this book is to support Advanced GCE Geography students in producing high quality fieldwork, whether at AS or A2. The table opposite summarises the requirements for each exam board. Clearly the first 'port of call' is for students to read about the precise **focus** for their chosen specification and to analyse the **mark scheme** thoroughly so they know what is expected of them.

The book is designed to be used on a **self study basis**. For each of the major areas for investigation in geography it provides:

▶ guidance on basic fieldwork techniques and enquiry work

▶ outlines of possible questions, hypotheses and issues for investigation by individuals **and** groups

▶ tips for success and how to avoid pitfalls which lead to major mark loss

▶ practical advice on key techniques such as sampling, questionnaire design and the use of statistics

▶ a range of ideas for representation and analysis of data

▶ specific risk assessment for any potentially hazardous fieldwork activities

▶ support for students in further secondary research using websites and other sources.

Both authors are experienced field workers and examiners. Their hope is that the practical nature of their advice will help students to enjoy fieldwork and to devise great personal satisfaction from the experience of producing individual enquiries to the highest possible standard.

Fieldwork safety

Fieldwork is generally a very safe activity, but all actions undertaken carry some element of danger or risk. Your job is to anticipate, minimise and manage any possible risks, starting at the planning stage and extending throughout the fieldwork. Use your common sense, be aware of your surroundings and follow the safety advice given below. Some of this will only be relevant in certain situations, but others will be applicable in all out-of-door activities. The advice given below applies to most fieldwork situations; more detailed comments relating to specific sites can be found in the relevant units of this book. Whatever you are doing, you should carry out a **risk assessment**. Safety remains your personal responsibility.

Examination Specification	Title of Work	Fieldwork AS	Fieldwork A2	Word limit	Subject area	Mark weighting for Advanced	C = centre assessed / E = externally assessed	Further Details
Edexcel A	Personal enquiry must include primary data collection	✓	optional	2500	Any branch of geography	20%	C	A skills paper alternative is available which requires students to have undertaken two days of fieldwork. Enquiry focus must be approved.
Edexcel B	Environmental investigation must include primary data collection	✓		2500	Linked to AS specification	16.6%	C	Can be a group or individual piece of fieldwork, but individual **focus** of final work essential.
	Research report		✓	1500	Linked to Unit 5B	7.5%	E	Can be primary and/or secondary research.
AQA A	Skills paper (includes investigative skills)	✓				15%	E	Candidates must have undertaken fieldwork.
	Fieldwork investigation (optional)		✓	4000	Any branch of geography	20%	C	Practical paper alternative whereby fieldwork is assessed via an exercise on a pre-released theme.
AQA B	Enquiry option	✓	✓	3500–4000	Any branch of geography	15%	E	Limited assessment of two days fieldwork requirement in Unit 4. Practical paper alternative.
OCR A	Geographical investigation (primary)	✓		1000		15%	E	Skills paper for assessment.
	Personal Investigative Study		✓	2500 or 1000		15%	E	Investigative skills exam alternative.
OCR B	Geographical investigation (primary)	✓		1000		20%	C	Skills paper for assessment.
	Investigation (primary or secondary)		✓	2500		15%	E	Either fieldwork or a research essay using IT.
Welsh (W) EC	Personal enquiry (geography assignment or individual investigation)	✓	✓	3500		15%	C	Investigative skills paper to assess fieldwork and other skills. One of two alternatives.

Safety issue	Managing safety (yours, the public and the environment)	Relevant areas
Clothing (hypothermia)	Fieldwork is miserable if you are cold and wet. Wear appropriate warm clothing and take water-proofs, hat and gloves when necessary. Your footwear should be fit for purpose and be comfortable.	All
Country Code	You must avoid fire, injury and stress to livestock and damage to yourselves, i.e. climbing loose stonewalls, crossing barbed-wire fences, avoiding bulls.	All countryside surveys
Drowning	When working at the coast, check tides and sea conditions. Crossing large rivers is very dangerous especially in flood. Immersion can induce hypothermia.	Coastal and river environments
Emergency kit	Try not to get lost (a map may help if you can use one). Make sure you know what to do in an emergency, i.e. where to summon help and what to do if you, or someone you are with gets injured. Consider taking a small first-aid kit, a torch and a whistle (and emergency food for remote areas).	All, excluding some urban environments
Herd mentality	Group work encourages a 'herd' mentality in which individual students switch-off from making their own decisions and lose awareness of the potential risks that surround them. It may also lead to anti-social behaviour, e.g. not considering other pedestrians on pavements.	All
Identification	When dealing with the public carry some form of personal identification as it makes other people feel safer with you, e.g. student card, laminated badge, letter from school/college.	All, especially questionnaire surveys
Maps	Absolutely vital – even in towns where a street map is essential.	All
Medication	If you regularly take/have medication do not forget it when you go outside, e.g. Ventolin for asthmatics or an Epi-pen for people with severe allergies. Suntan lotion is also important when it is sunny.	All
Missing	Always tell a responsible adult where you are going and when you are expected back. Carry a public transport timetable if appropriate. If possible take a mobile phone, but don't rely on it to work everywhere, especially in remote locations.	All
Money	Always take some money with you (including change and/or a phonecard for the telephone). You may need it to get home in an emergency or to buy an ice-cream!	All
Paths and tracks	Never run down hills as things can go out of control very easily. Take care on steep paths and tread carefully on potentially slippery ground, especially wet grassy banks, boulders in rivers and on the foreshore.	Mountainous and rural locations
Permission	Make sure you obtain written permission to enter private property (including shopping malls).	Anywhere private
Traffic	Be sensible, be aware of your surroundings and do not expect drivers to be aware of your presence, especially in quiet rural areas.	Towns and country lanes
Working alone	In most situations it is not advisable to work on your own. Bring a responsible friend or relative who can help.	All, except some urban situations

Fieldwork management

Research and choose title (may arise out of group fieldwork)

Fill in proposal form after rigorous pre-research to ensure viability

Interview 1 with teacher

Agree on the title

Approved by Moderator

- Research sites for data collection
- Phone all key sources of data to ensure you can do the project

- Research methods from fieldwork books
- Look at formats of previous projects which did well. Enquire whether there are any secondary data which you might use from these projects.

- Pre-pilot questionnaires
- Design sampling methods
- Devise booking sheets

- Carry out primary research (three days for large A2 projects)
- Collect together all secondary research sources

Discuss your findings and whether your data collection is OK and what else you need to do.

Interview 2 with teacher

- Keep a diary of all your visits, phone calls as evidence, as well as your raw data.
- Take photos of yourself doing the fieldwork

- Review aims
- Get all diagrams, maps and graphs drawn
- Write a rough draft on what your findings show (including table of research methodology)
- Present all your spreadsheets
- Organise data for statistical analysis
- Consider possible conclusion

- Look at wider significance
- Research on Internet
- Look at fieldwork books to consider means of presentation
- Find out about word limit and any penalties
- See how diagrams can save you words
- Process statistics

- Discuss presentation
- Write up
- Record any deficiencies
- Discuss conclusions

Interview 3 with teacher

- Check computer systems and availability of workstation
- Decide on method of binding. Check regulations on folder type etc.

- Word process final product. Make sure you obtain the mark scheme to decide on format.
- Check photographs and maps are presented well.
- Check all figures are numbered
- Include bibliography and acknowledgements

The finished product

The precise format of your project will depend on your exam board. Most exam boards publish project guides and mark schemes so be sure to use these. Looking at successful projects of former students is also a good way to see how you can obtain a high mark.

Getting started

▶ Carry out a personal audit identifying your potential strengths.
▶ Find a topic (local newspapers and past projects can help).
▶ Think about a possible question to research.
▶ Research and map your area and look for survey sites. Decide on a sampling strategy.
▶ Identify possible secondary sources (e.g. library).
▶ Formulate your title and research programme.
▶ Carry out a risk assessment.

Setting the scene

▶ This can be done before you do your fieldwork. You need to finalise your aims and justify any hypotheses.
▶ Refer to past research on the topic (old projects, historic data).
▶ Research the theory of the topic.
▶ Set the scene for the project. Obtain a map and aerial photo of the study area. Investigate the main issues and give a brief history of the study area. www.multimap.co.uk is a useful site.

Methodology

▶ Pilot and develop questionnaire(s). Develop booking sheets ready for database.
▶ Work out sampling strategies and summarise the rationale.
▶ Check availability of specialist equipment from your centre.
▶ Find a companion to work with – check all health and safety issues.
▶ Organise any group research.
▶ Identify what you can use from group research.
▶ Collect the data and bring back to be checked. Research secondary sources.
▶ Evaluate and criticise data sources/collection.

Results analysis

▶ Process all the data, e.g. provide summary spreadsheets of questionnaires and tables.
▶ Get all the diagrams, maps and graphs drawn. Make sure your maps all have scales and direction arrows, and your graphs carry a title, and a footnote with the information source.
▶ Organise all your photographs and sketches to use them to the best advantage.
▶ Incorporate all the data from secondary sources and acknowledge.
▶ Decide what statistical techniques you are going to use and process them taking care that you interpret them rigorously.
▶ Assemble all your diagrams and describe and analyse what they show. How do the results measure up to expectations?
▶ Look at values analysis by developing a conflict matrix.

Evaluation and conclusion

▶ Recall and summarise your main findings.
▶ Evaluate the success of your project, commenting on the wider significance of your conclusions. Are they similar to published research? Do they show change?
▶ Include a section on limitations of your enquiry.
▶ Write a brief summary of further research you could carry out.

Presentation and general advice

▶ Organise your project under the mark scheme headings for your specification. Number all the figures and pages.
▶ Organise the appendix to include evidence/diary of data collection (raw data) supporting resources.
▶ Include bibliography of sources and acknowledgements of assistance given.
▶ Word process your project as this makes it much easier to reorder, to cut out words etc.
▶ Keep to the word limit (use tables, diagrams etc.) but do not cut short on the analysis and conclusion.
▶ Make sure you do some good hand drawn maps as well as computer drawn maps.

1 SKILLS

SAMPLING
Advantages of sampling
Sampling methods
Evaluation of sampling techniques

QUESTIONNAIRES
Preparing your questionnaire
Delivering your questionnaire
Quality indexes
Qualitative techniques

MAPPING

GRAPHICAL SKILLS

USING PHOTOGRAPHS

Skills

SAMPLING

Although in some cases you can survey a whole **population**, for example a small hamlet of about 50 people, in most cases it is not possible so you have to **sample**.

Advantages of sampling

Sampling is **quicker** and therefore cheaper (for a commercial project). For some projects such as measuring pebbles on a beach, the erosion levels of footpaths, or the nature of an ecosystem such as a woodland, it is **impossible** to measure everything because the population size is too great. Once you start measuring or recording you will find a pattern emerges, so it is **unnecessary** to measure the whole population. Sometimes when you are doing questionnaire work you get **refusals** so you cannot survey the whole population. In some cases you do not know the whole population, for example how many shoppers are in town that day. In some cases you need an instant result (known as a **snapshot survey**) so you have to sample to get coverage – for example when doing a pedestrian survey of a busy town at key times throughout the day.

From your sample you can **estimate** what is happening in the whole population, provided you get the technique correct and avoid **bias**. Rigorous sampling is very important and you must not make sweeping conclusions from limited and biased data.

▶ **Design the sampling frame correctly.** For example it is much better to use an electoral register for selecting addresses than a telephone directory (as you may exclude some lower income groups and therefore introduce bias).

▶ Care with choice of sample is vital. For example if you are looking at consumer choice (such as coffee/tea drinking) you need to ensure you get a balance of age, gender and any other relevant factors. This is always hard as often your sample is biased towards old ladies, who do not work and are very willing to talk. If you are doing pedestrian counts or traffic counts you need peak/non peak times on different days.

▶ You need to avoid very small samples (except for in depth perception surveys). 50 is usually a sensible minimum, but for a short catchment area questionnaire, 200 is a more useful target.

Sampling methods

There are four reconnaissance sampling methods you can choose from.

Figure 1.1 The four reconnaissance sampling methods

1. Point sampling
This involves choosing a number of individual points, e.g. houses on an estate, streets in a district, or fields on a land use map.

2. Grid sampling
A quadrat is an area (usually a square area). The size depends on what you are sampling – usually 0.5m² or 1m² for vegetation, beach or river deposits but up to 100m² for a woodland.

3. Line sampling
Line sampling involves taking measurements along a line, for example looking at processes on a slope, micro-climate measurements or pollution measurements across a city. It is usual to sample at a regular interval, e.g. every 10m for detailed measurement.

4. Belt transect
This method involves a survey of a continuous area about 1–2m in width. It is ideal for looking at a succession on a rocky foreshore, or across a sand dune.

There are three in-depth sampling methods as shown below.

Figure 1.2 The three in-depth sampling methods

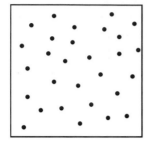

1. Random sampling

Random sampling has a specific **technical** meaning. It is not shutting your eyes and stabbing at a map with a pencil, or throwing a quadrat over your left shoulder. It means sampling using random numbers.

▶ Decide on how many points, quadrats or lines you want.
▶ Grid out your area with an overlay and number the lines.
▶ Use random number tables to generate the points on the map (or the two ends of the line).
▶ Visit the sites to do the survey work.

You can use random number tables to decide on house visits or people to interview.

Advantages No human bias in selection.

Disadvantages The points may not cover all parts of a varied area, e.g. 38/50 might be in a council estate which covers 10 per cent of the ward.

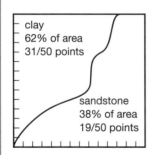

clay
62% of area
31/50 points

sandstone
38% of area
19/50 points

2. Stratified sampling

Here you design your sample to include a representative proportion of the sample, for example of land use on two rock types. Plot the points as for random sampling, but having decided what per cent you need for each ignore any points over 19 for sandstone, and keep on with clay until you reach 50.

Advantages It reduces bias arising in an area of contrasts.

Disadvantages It is difficult to get the stratification correct.

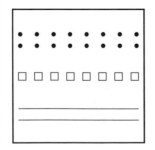

3. Systematic sampling

Here the sample is chosen according to an agreed interval, e.g. every fifth house or person, or every 50m on a line, or by placing lines, or quadrats at regular intervals.

Advantages It provides more complete coverage of an area. It is quick and there is no need to bother with random tables.

Disadvantages The points or lines may miss variations and result in bias. Linear features such as a main road may be missed.

Evaluation of sampling techniques

You must always ensure that in your methodology section you write a **justification** of your choice of sampling method and technique and also assess any problems that arise during your sampling.

You can also evaluate the size of your sample (try to achieve a minimum of 30 for quadrats or complex questionnaires and 50 for standard questionnaires). You need to know how far the mean of your reading deviates from the true value. If your survey involves numerical measurement use the following formula.

$$\text{Standard error of the mean (Sm)} = \frac{\text{S (standard deviation)}}{\text{n (number of samples)}}$$

If your data consists of proportions, e.g. per cent vegetation cover, use the following formula.

$$\text{Standard error of the mean (Sm)} = \frac{\text{P (\% of land in a particular category)}}{\text{N (number of points in sample)}} \times \text{q (\% of land not in category)}$$

Note that standard error of the mean goes down as sample size increases.

To find out whether you have got the right size of sample and whether your sample is a true estimate of the characteristics of the whole population, you need to follow the following steps.

▶ Calculate the standard deviation for your sample (s).

▶ Decide on the tolerable margin of error (d).

▶ Decide on the required level of certainty (z) e.g. 90 per cent.

▶ Calculate the required sample size $\frac{(z \times s)^2}{d}$

QUESTIONNAIRES

Questionnaires are the most widely used primary data source, especially for human geography projects. They enable you to collect **activity** data so you can find out what people do, how often and when, and also **attitudinal** data so you can find out people's opinions. You also need to collect background or classification data so you can analyse the results from your sample, e.g. age/sex/location of residence etc.

Questionnaires are unique in that you can find out information about people's opinions and behaviour (for example personal decisions about shopping) which are not available anywhere else. The

information is also up to date and can be compared with historical data. Questionnaires however have many disadvantages. In general people do not like answering questions so you have to be prepared for rudeness. You often get a very biased sample as some people do not have time to stop, and others (usually old ladies) are very willing to answer questions. People give weird answers or tell you what they think you want to hear.

Preparing your questionnaire

Writing questionnaires is **not** easy. Figure 1.3 shows an example of a satisfactory questionnaire, and one which would have been a disaster.

The following points need to be considered when planning a questionnaire.

▶ Keep the questionnaire as **short** as possible. Every question must be of direct use, for example if you only want catchment area data you only need to ask where people come from (location/postcode). However, you cannot go back. A maximum length is probably about eight to ten questions.

▶ Carry out a **pilot** with family and friends to check the questions are clear and unambiguous and that you are not missing out important issues.

▶ Do not include any questions which can be answered elsewhere.

Figure 1.3 A satisfactory questionnaire and a poor questionnaire

Include a first sentence to help you approach people. Don't forget to explain why you want their help.

Pilot showed it was difficult to use boxes because it was so variable

Best to be open ended then can be classified.

Ideal for statistical analysis but you can only find time to ask one of these.

Delete – not needed.

Needs to be open ended but 2 areas difficult.

Delete

Avoid asking personal questions – make a judgement yourself afterwards.

Always start with non-sensitive factual questions as a 'warm-up'. Leave more difficult probing question to end.

'Closed' questions will help your processing and analysis of data.

Open questions help you find out more about people's opinions, but leave them until last once people are confident with you.

A satisfactory questionnaire

Good morning/afternoon. I am a sixth form student from Rotterdam investegating the impact visitors have in the Malham area as part of my Geography project. Could you spare me a few moments to answer some questions?

1. a) Where do you live, which town/village do you live in (post pilot change)?
   b) How far have you travelled to get here?

2. How did you get here?
   - Car ☐
   - Walk ☐
   - Coach ☐
   - Bus ☐
   - Bicycle ☐
   - Motorbike ☐
   - Tram ☐
   - Other

3. How long are you staying in the area for?
   - Half a day ☐
   - 2–7 days ☐
   - 1 day ☐
   - > 7 days ☐

4. How will you spend your time here? (tick as many boxes as required)
   - Walking ☐
   - Picnicking ☐
   - Cycling ☐
   - Caving ☐
   - Rock-climbing ☐
   - Running ☐
   - In the pub/cafe ☐
   - Other ☐

5. How often do you visit the area?

6. What attracts you to this area?

7. To what extent do you think that visitors create the following in the Malham area? (0 = not at all. 5 = significant impact)

Parking problems	0	1	2	3	4	5
Litter	0	1	2	3	4	5
Footpath erosion	0	1	2	3	4	5
Damage to vegetation	0	1	2	3	4	5
Traffic problems	0	1	2	3	4	5
Vandulsiam and graffiti	0	1	2	3	4	5
Employment opportunities	0	1	2	3	4	5
Economic benefits	0	1	2	3	4	5
Other	0	1	2	3	4	5

8. Overall, how would you rate the impact of visitors to the area?

9. Do you think the area is managed effectively by the Yorkshire Dales National Park?

10. Are there any suggestions you could make for improvements in the park area in the village?

11. Any other comments?

Estimate of age
<20 ☐   21–35 ☐   36–50 ☐   51–65 ☐   >66 ☐
Male ☐   Female ☐

No context

Personal sensitive data

Biased leading question

Very closed yes or no

Difficult, needs structuring

Unrelated

A poor questionnaire

1. What sex are you?   female   male

2. Do you use the new by-pass: weekends   weekdays

3. Do you agree that the by-pass has increased trade in the town centre and encouraged the development of industry on the edge of the town?

4. Have you noticed an increase in the volume of traffic in the town over the last five years?

5. Did you know that the new by-pass cost millions of pounds to construct?

6. Do you think that building the new by-pass was money well spent?

7. What other impacts do you think the by-pass construction has had on the town?

8. What has been the impact of counter-urbanisation on the town?

Make sure there are no leading questions as you need to avoid bias.

The sample size for your questionnaire is important if your results are going to have any validity. Clearly the larger the sample the better. For a standard length questionnaire, 50 is reasonable but for a questionnaire to delimit a catchment area 200 may be the minimum.

You need to give thought to the sampling **method**. For instance if you are thinking about a stratified sample you need to look at the whole population first, and then decide on proportions by age, gender, income etc.

Delivering your questionnaire

1. Stand in the street and catch passers by. The site chosen, the day of the week and the time of day are very significant. For some sites, e.g. shopping malls or superstore car parks, you must get permission to do surveys.

2. Go from house to house in order to cover a cross section of housing types in the Enumeration District or Ward.

3. Post the questionnaire to specific houses and collect them later or enclose an SAE.

For both (1) and (2) you should always work in pairs and choose a safe time of day.

Method number 3 is quicker but it is rare to get more than 30 per cent take up. Make sure you enclose a polite letter with the questionnaire and your school address.

If you use (1) or (2) you may devise a large booking form to log responses rather than individual questionnaire forms, as this saves processing time.

Quality indexes – an alternative to questionnaires

Many students struggle to collect enough primary data without resorting to endless questionnaires. Quality indexes are useful because they rely on observation and can provide data for statistical or graphical analysis. However, these indexes usually rely on qualitative judgements made by the observer. Photographs are vital to provide evidence as to how judgements are arrived at.

There are a number of basic indexes which can be used in a variety of projects. These are shown below with some suggestions for use. In some cases (e.g. index of decay) an ordinal scale is used but in other examples (e.g. the environmental street index) a biopolar scale is used to look at good or bad points.

	Index of housing decay		Table 1.1		
For each house chosen	**None**	**Little**	**Some**	**Much**
Deterioration of walls	0	1	3	5
Paint peeling	0	1	2	3
Displaced roof material	0	1	5	9
Broken glass in windows	0	1	3	7
Broken gutters, etc.	0	1	3	7
Structural damage, e.g. settling cracks	0	3	6	11
Rotting timber	0	2	4	8
Sagging roof	0	2	6	10

Either in the field if time, or on return to school for every street examined, add together the awarded points then subtract your total from 60.

The following general points can be made from your results.

Score	Physical condition of buildings
50–60	Good/excellent
40–49	Satisfactory
30–39	Generally unsatisfactory. May be bad in specific points
20–29	Action needed in very near future to improve structure
Below 20	Need to demolish or rebuild

Index of housing condition (decay)

Useful for house price surveys and quality of life surveys. Can be combined with street quality to give an overall environmental profile, which can be used as an indicator of social class.

Landscape quality surveys

Especially useful to relate to activity surveys/visitor density surveys, or as a basis for impact surveys.

Landscape quality Table 1.2

	Score	
	good **bad** +3 +2 +1 0 –1 –2 –3	
Interesting		Boring
Attractive		Unattractive
Varied		Monotonous
Like		Dislike
Welcoming		Hostile
Historic		Modern
Well maintained		Neglected
Open		Enclosed

Street survey Table 1.3

Feature	Penalty points	Maximum score	Feature	Penalty points	Maximum score
Landscape quality Trees and well-kept grassed spaces Few trees and/or unkept grassed spaces No trees or grassed spaces	 0 4 8	 8	**Noise** Normal residential standard – quiet Above residential standard – with some noise Main street standard – very noisy	 0 2 5	 5
Derelict land None Small area Large area – a major eyesore	 0 4 10	 10	**Air pollution** No offensive smells or obvious air pollution Offensive smells and/or obvious air pollution	 0 10	 10
Litter/vandalism No litter: no vandalism Some litter or vandalism Very untidy: much vandalism	 0 4 8	 8	**Access to public open space** Access to park/public open space within 5 mins. (500 m) walk No park/public open space within 5 mins (500 m) walk	 0 3	 3
Industrial premises All residential properties Some industrial premises Mainly industrial premises	 0 5 10	 10	**Access to shops and primary school** Primary school and shops within 5 mins (500 m) walk Primary school only within 5 mins (500 m) Shops only within 5 mins (500 m) walk No primary school or shops within 5 mins (500 m) walk	 0 2 3 5	 5
Traffic flow Normal residential traffic Above normal residential traffic Heavy vehicles and through traffic	 0 3 6	 6			
			Maximum Penalty Points =		65

Street surveys

Useful for determining voting habits, quality of life surveys, incidence of crime, or as an indicator when looking at the impact of regeneration etc.

	Shopping quality	Score (on a scale of 1 to 5)
A	Type of shop	1 = dominated by department/variety stores or shops selling comparison goods
		5 = wide variety of shop types, convenience goods dominant
B	Other land-use groups	1 = mainly shops
		2 = shops and banks/building societies
		4 = mainly offices
		5 = very few shops – dominated by houses/industry
C	Retail organisations	1 = national chain stores dominant
		3 = mixed – some national and independent
		5 = small independent shop units
D	Quality of goods	1 = good quality and/or high price goods
		5 = low quality and/or low price goods

Shopping quality surveys (Table 1.4)

Very useful for comparing CBDs, looking at zoning, comparing the impact of pedestrianisation, or the development of out of town stores.

Other indices can be devised for more specialist purposes.

Quality of services (Table 1.5)

The index shown in Table 1.5 is for a cinema, but you could devise your own for a leisure centre, swimming pool etc. You need to see what is important, e.g. value for money, size, facilities, cleanliness of changing rooms, safety etc. may all be important for a swimming pool. You can use the quality survey when considering the adequacy of the service creating a service hierarchy, or linking the service to a catchment area.

Land capability index (Table 1.6)

This is one of a number of indexes which are based on actual measurements, e.g. biotic indices to measure water pollution. Land capability is a vital index for relating land to potential land use. It is also very important when considering land loss for new developments, e.g. golf courses or housing on greenfield sites.

CINEMA QUALITY INDEX						Site 1 The Roxy
Bad features	1	2	3	4	5	**Good features**
Old films			✓			Latest films
Uncomfortable seats		✓				Comfortable seats
Dirty	✓					Clean
Difficult to get to	✓					Easy to get to
Expensive				✓		Cheap
Sub-Totals	2	2	3	4	0	Total = 11

A land capability index | Table 1.6

Class	Altitude (m)	Wetness	Soil quality	Soil fertility (ph)	Slope (°)
1 High quality	Below 100	No limitations Free drainage Rainfall <750 mm	Deep soil 75 cm + Stone free Loam texture	7 + (neutral)	Level (not above 3)
2	100–150	Imperfectly drained Drainage easily modified by liming	Depth 50–75 cm Slightly stony	6.0–6.5	Slight (not above 7)
3	150–200	Some problems but possible to install drainage system	Depth 25–50 cm Stony – may be sandy or clayey texture	5.5–6.0	Moderate (not above 11)
4	200–350	Poorly drained but can be improved to maintain pasture	Shallow – under 25 cm Very stony	5.0	Significant (11–20)
5 Low quality	Above 350	Poorly drained Drainage almost impossible to install Rainfall > 1,250 mm	No humus Very stony – skeletal soil only	Under 4.5	Steep (over 20)

Quality of site analysis

This type of analysis looks at good and bad siting factors for e.g. a village and then adds extra points for situation. You can build up an index of centrality based on the number and range of services and relate this to the site index and the amount of growth. Other examples of site indexes include office sites (design of building and neighbourhood) or industrial obsolescence where you can compare old and new industrial sites.

Using this **index**, you can evaluate a site – assess how good it is (or isn't).

Award marks – from 1 to 5 – for the features mentioned. Then add them up.

A mark less than 30 means a poor site; a mark above 30 means a good site.

A quality of site analysis | Table 1.7

Disadvantages	1	2	3	4	5	Advantages
steep relief						flat land
on high land						near sea level
faces N–W						faces S–E
prone to flooding						not prone to flooding
far from natural water supply						close to natural water supply
thin soil, clay or sand						deep soil, loam
no humus						humus >12%
soil pH >8 or <4						soil pH 6.5
impermeable bedrock						permeable bedrock
no room for expansion						room for expansion
SUB-TOTALS						TOTAL POINTS

Quality indexes can be used in a large number of ways as shown below.

	Possible criteria	Use in a project
Car parks	Quality of surface, ease of access, nearness to town centre, size of space, safety of car (security), overall size, cost.	Traffic management or CBD Survey.
Beach quality	Swimming conditions (calmness/safety for children), quality of sand (freedom from stones etc.), facilities, cleanliness of water, litter, access, shelter.	Use with landscape quality surveys to assess variations in usage. Need for activity surveys.
Bus/train services	Quality of vehicle, capacity/seating, cost, route, reliability/punctuality, frequency.	Usage and provision of surveys, e.g. before and after deregulation of buses or preservation of a rural branch line.
Footpath state	Width of footpath (freedom from weeds), litter, dog dirt, degree of erosion (width/depth), infiltration/compaction rates, surface type.	Relate to usage. Survey of availability. Exploring problems of mountain biking, horses, way marking, erosion.
Road condition	Surface (pothole survey), drainage, whitelining/cats eyes, litter etc. Saturation index (vehicles per hour). Density index.	Relating to problems of finance.
Mobility index	Kerb height, pavement width, pedestrian density, smoothness of surface (cobbles), traffic flow v pedestrian crossing.	Useful for a disabled provision survey in a CBD, can be combined with provision of shops and services.

Qualitative techniques

There are a number of other interview techniques you can use in what is called action research. They are all very useful for certain types of qualitative projects. They also help you to achieve Key Skills Qualifications.

For some projects a number of in-depth interviews will be extremely important. For most projects a key interview with a planner or an industrialist is vital so you can get an expert opinion or explore the background to an important location decision (for example company policies for superstore location).

Interviews should be planned. Seek guidance from your teacher in composing a suitable letter requesting an interview. You should indicate some of the key issues you want to explore. Make it clear what use will be made of the results, and indicate the likely length of the interview. Write a formal thank-you after the interview.

You need to decide on a number of questions, preferably open ended, so you can understand the interviewee's perspective on the issue. It is often useful to ask permission to record the interview so you can make a transcript for analysis. You may decide to interview a number of key players involved in a controversial issue and prepare a values analysis so you can develop a conflict matrix. Textual analysis of various documents in a planning enquiry can support the development of a conflict matrix.

Oral histories can be very useful for evidence of change, for example in holiday habits, or changing villages or farming practices. They can also be very useful for demographic studies such as population structure, or health. Ideally you need to interview about ten people. They will of course tend to be elderly because of the historical slant, but you may decide to go for a time slice looking at change over decades. You need to ask open ended questions but within a pre-designed structure. Such interviews should take around 20–30 minutes and you may be able to produce annotated diagram summaries. Ask permission to use a tape recorder so you can summarise a transcript of the interview.

Focus groups can also be very successful. A number of people are assembled for a discussion concentrating on a particular topic, such as town centre regeneration. In some cases it may be feasible for you to use a pre-existing group, and seek their opinions about a related issue. If for example you are looking at recreation provision, you could assemble a group of junior school children, a group of teenagers, or visit a pensioners club. You need about 6–12 members, and at least 30 minutes of discussion. You, as the researcher, have to take control, so you will need to record the discussion (preferably by video). You can then summarise opinions.

Participant observation (for example when you are doing an activity survey on a beach, shopping centre, or in a honeypot site) can be augmented by an interview which can be used to construct a space-time diary. You can ask sample visitors what time they arrived, what they did (e.g. 9.30 – 10.30 sat by car in car park) and where they went on their visit. Equally, at a retail park or shopping centre you could ask when they arrived, what stores they visited etc. to build up a usage profile. This is perhaps better than **critical path analysis**, where you follow selected 'victims' through a shopping centre or in a park to record how they spend their time.

Perception studies can be an interesting supporting technique, for example in an enquiry on house prices where you survey the perceived ranking order of a number of neighbourhoods for their desirability as a living environment, or alternatively how people perceive risks, for example in a coastal erosion or flood hazard survey. The perceived risks can be compared with reality. The results can be recorded on

a bipolar scale, and related to income, location and age/gender. If you are doing a crime project it is an excellent idea to develop a map of fear, by showing residents photographs of local areas, and developing a scale of safety/unsafety depending on whether they see areas as likely scenes for mugging. The incidence of burglar alarms is a good indicator of the perceived risks from crime and can be related to police or newspaper data of actual crime incidents.

You may decide to do a project which relies heavily or exclusively on perception studies. For example, in order to ascertain whether the CBD of a particular town has become too diffuse, you could duplicate a set of base maps and ask a survey group of 20 to mark where they considered the core, the frame and the fringe of the CBD to be. A study based entirely on perception is one on **mental maps** and the factors which influence the nature of them, such as age, gender, occupation, length of domicile, ethnicity, place of habitation and disability.

You structure your sample, get the respondents to draw their mental maps and then analyse them in terms of routes or paths, edges, nodes (meeting points) and districts (neighbourhoods they clearly identify). You can study how people perceive distance, both in metres and journey time. This can come in useful when assessing the quality and provision of car parks. Perception studies are sometimes used by bodies such as National Parks, when considering visitor management. In order to take pressure off a number of key 'honey pot' sites, they have to create alternatives.

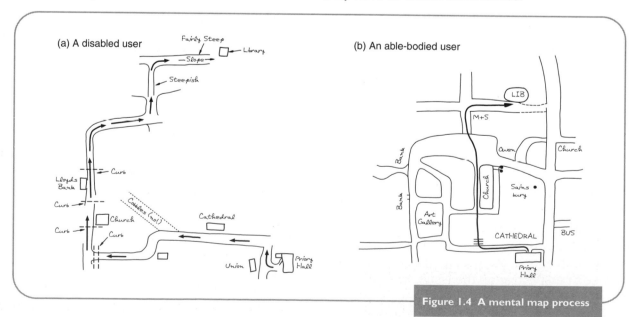

Figure 1.4 A mental map process

MAPPING

Well drawn, accurate maps can add a great deal to a project. In most cases it is very straightforward to hand draw your own base map and then photocopy it for **all** your needs. Some of the main mapping methods available are shown below, together with suggestions for their use. Always remember to include a title, a scale, a north direction and a key to your maps.

Dot map

Data may be shown on a map by drawing dots to illustrate the distribution of the items, e.g. post offices, people, earthquakes. If the number of points is small and their exact location is known the dots are drawn and the pattern of dots can be statistically analysed using **Nearest Neighbour analysis**. You can also draw different sized dots to compare relative importance.

There are three points to consider:

Dot values – decide how many dots you want and divide this into the total number of items to get a value.

Dot size – can be used to generate a hierarchy, e.g. based on size of a school or town.

Dot location – if information is to represent areas, not locations, distribute dots evenly.

Dot maps can be used for pedestrian counts, tourists on a beach/honeypot or in a park. They are very useful for showing densities of people or animals within defined areas, e.g. wards/parishes.

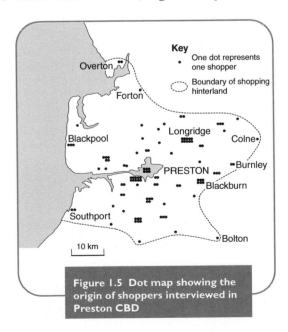

Figure 1.5 Dot map showing the origin of shoppers interviewed in Preston CBD

Choropleth

On choropleth maps areas are shaded according to a key. Each colour shading represents a range of values. You should use a dispersion diagram to split your data into four equal groups or you can develop categories such as 0–9.9, 10–19.9 etc. at equal intervals (but the groups may not be equal). The shading should be graduated from dark/dense for high values to light for low values. Do not forget to draw in the key. These maps are good visually as they show high/lows but are not good for accurate values and give a false impression of abrupt change at boundaries.

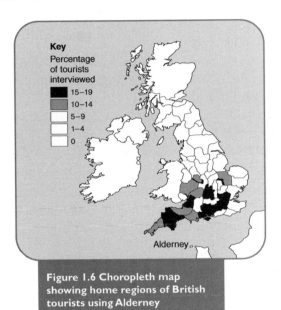

Figure 1.6 Choropleth map showing home regions of British tourists using Alderney

Isoline map

Isoline maps show lines of equal value. They can be used when the variable to be plotted changes gradually across an area and where plenty of data are available.

▶ **Plot** the data as a series of points with the values written beside each point.

▶ Decide on an interval.

▶ Sketch the isolines for the chosen interval using curved lines. For lines drawn between points you will have to estimate.

▶ An isoline never splits.

▶ An isoline should never be missed out even if the rate of change is steep.

Isoline maps show changes over an area but are **unsuitable** for patchy data. They are subjective as two people can get different interpretations from the same data.

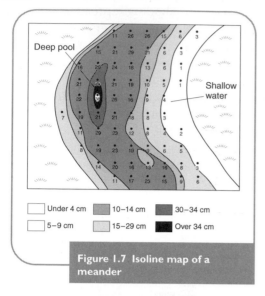

Under 4 cm 10–14 cm 30–34 cm
5–9 cm 15–29 cm Over 34 cm

Figure 1.7 Isoline map of a meander

GRAPHICAL SKILLS

Many types of graph can be produced to a high quality on a computer, using a graphics package such as Excel, rather than hand drawing them. You should show your graphicacy by hand drawing some of the types of graph not available in the package, or in innovative types you develop yourself. You need to be very careful about choosing an **appropriate** design for your graph on the computer as it is easy to get carried away by the varied menu. Data need to be input accurately and in sufficient quantity to produce a reliable result.

Line graphs are used when measurements are **continuous**, e.g. temperature, traffic movement, river flow, employment trends. They show the relationship between two variables, one of which causes the other to change, e.g. temperature with distance from the sea. Distance from the sea causes the temperature to change so it is the independent variable (horizontal or x axis) with temperature as the dependent variable (vertical or y axis). Always start axes at zero and choose a suitable **scale** or your graph will give you a distorted visual picture. Always label both axes. You can develop multiple line graphs (to compare sites) or comparison line graphs to show different activities over the same time period. If you have very large ranges of results and you find it difficult to fit the results in, you may need to use **semi log** or **log-log** graph paper.

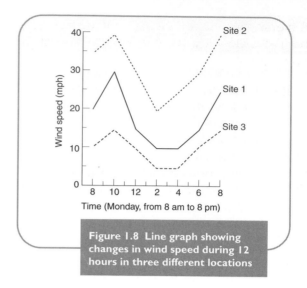

Figure 1.8 Line graph showing changes in wind speed during 12 hours in three different locations

Bar charts You need to make sure you are clear about the difference between a basic bar chart and a histogram as they do look very similar. In the case of bar charts the horizontal scale represents a number of categories and the vertical scale the values they represent. The width of the bar is not important and often a gap is left between bars. In the case of a histogram the horizontal axis represents a linked series of class **intervals** e.g. pebble size in mm (0–3, 4–6, 7–9). Equal intervals are usual as the **area** of each bar is proportional to the class frequency. Many variations of bar charts exist.

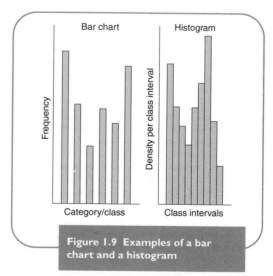

Figure 1.9 Examples of a bar chart and a histogram

▶ **Mirror bars** are drawn back to back to compare findings (e.g. frequency of plant species at two sites).

▶ **Population pyramids** are used to show population structure.

- **Reverse bars** show increases and decreases (the bar is drawn below the line) for example of population.

- **Divided bars** can be used to show relationships such as relating land use to geology.

- **Composite bars** can be used to show the proportion of each component category either to a common scale, or in the case of **proportional** bars the height of each bar is drawn at the correct length and then components calculated as a per cent.

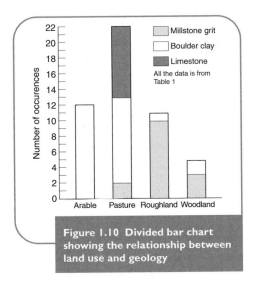

Figure 1.10 Divided bar chart showing the relationship between land use and geology

Pie charts are circles which have been divided into sectors whose area is proportional to the size of the component categories in a given sample. **Conventionally** the largest category is drawn first and progress is made clockwise from the 12 o'clock line in descending order. As with bar charts, mini pies displayed on a map or cross section are very effective.

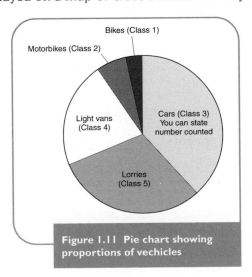

Figure 1.11 Pie chart showing proportions of vechicles

Scattergraphs can be used to explore the relationship between two variables, provided you have data for lots of places. The pattern of the scatter describes the relationship. You should draw a best fit line to show the trend (there must be an equal number of points above and below the line you draw in). You can calculate the position of this line by **regression analysis**. Your graph may show a positive (direct) or negative (inverse) relationship, and may also show anomalies which you will need to explain. If the relationship is not clear you will need to use a statistical correlation technique to see whether there is a significant relationship.

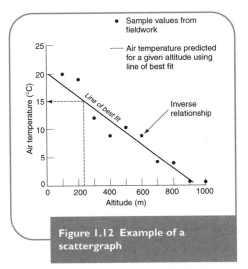

Figure 1.12 Example of a scattergraph

It is also possible to use a similar method to show the connection between two variables where only one has a quantitative value using a dispersion diagram variant which relates house type to price or land use to slope, altitude, soil pH, etc.

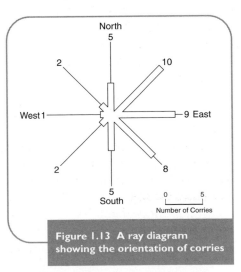

Figure 1.13 A ray diagram showing the orientation of corries

Circular or rose diagrams are used when the observed data takes the form of a compass direction. There are many ways in constructing these diagrams (reflected, half and single roses). These graphs can show wind direction, pebble orientation, or glacial features orientation. The example shown is for a number of corries and shows a simple rose diagram. You can get many interesting varieties of circular graph using a computer but you have to be very certain of what you are trying to achieve.

Pictograms (often used on advertisements because they are very eye catching) are less accurate than other graphs. All sorts of symbols (again available on the computer) can be used. They are very good for when you present a questionnaire as they do add variety.

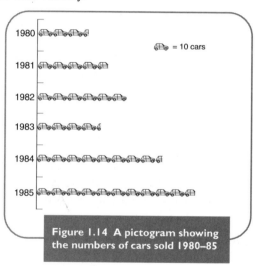

Figure 1.14 A pictogram showing the numbers of cars sold 1980–85

Located bar charts: one useful way of displaying results is to actually locate your bar charts on a map of where the readings were taken. You can make the

map even more sophisticated by using **proportional** symbols, i.e. the symbols are drawn on maps proportional in size to the variable being represented. Most common symbols are circles, bars, squares, and occasionally spheres or cubes.

Proportional bars

❯ Look at your data and decide on a scale. If the bars are too long they get tied up, and if they are too short the bars are difficult to read.

❯ Draw the bars on your base map, the bottom end is usually located next to the place it refers to.

❯ Bars should be of uniform width, solid and they are usually drawn vertically.

Proportional circles
These can be used for very wide ranging data.

❯ Convert the data into square roots.

❯ Convert the values into mm to give the radius of each circle.

❯ Draw the circles locating them in the centre of each area they refer to.

❯ Mark the scale on the map. These work well but are hard to draw neatly. Proportional squares are also a possibility.

Flow line maps are used to show **flows** or movements along a given route, e.g. traffic flows, migrations. A line is drawn along the route proportional to the volume of flow.

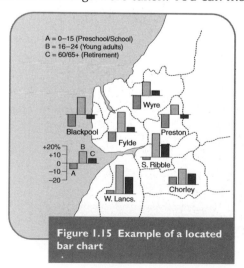

Figure 1.15 Example of a located bar chart

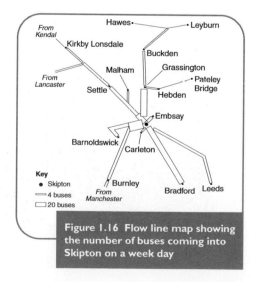

Figure 1.16 Flow line map showing the number of buses coming into Skipton on a week day

- Draw a base map of route surveys.
- Mark in points where data were recorded and their values.
- Decide on the scale or width of the line e.g. 1mm for 100 cars per 30 minutes.
- Use square roots of numbers if the lines come out too thick.
- Draw flow lines (these can be straight or actual routes).
- Draw a scale so the thickness can be measured.

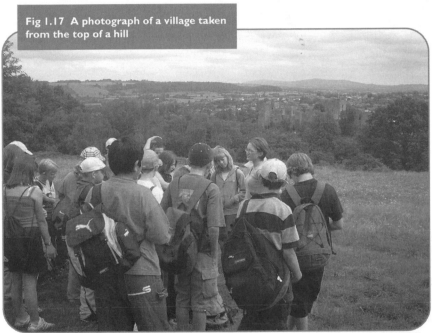

Fig 1.17 A photograph of a village taken from the top of a hill

USING PHOTOGRAPHS

Photographs can be a real bonus to your fieldwork project, but you need to think about how you can make the best use of them. You need to have a clear vision of how they fit into your fieldwork before you take them. Always record the precise location (grid reference), time of day, weather conditions and the direction the photograph is taken in. Annotate your photograph to bring out the points it is supposed to be making. This can be done by using an overlay of acetate or you can 'scan in' the photographs.

1. Photographs (especially commercial aerial photographs) can be very useful for introducing the main issues you are going to investigate. Figure 1.17 showing a carefully taken semi aerial view (from the top of a hill looking down) showing the site and layout of the village.

2. Photographs can be very useful to give a visual display of the issue you are investigating, for example, footpath erosion, eutrophication in a stream or weather conditions. They can show change over time, for example, stratification in a woodland or the changes in pedestrian/traffic flows in a tourist town. You can also use old photographs to show **change**.

Figure 1.18 A photograph of deciduous woodland in a) summer, and b) winter

3. Photography can be used to show how you undertook surveys or field measurements. You can get a friend to take a picture of you in action, for example measuring the girth of a tree or interviewing a member of the public. You can also show how you used equipment, for example to measure the velocity in a stream. Photographs of the equipment used and its limitations are also helpful, for instance if you made some homemade weather measuring equipment.

Fig 1.19 A photograph of students measuring the velocity of a stream

4. Photographs are very helpful for showing key features, for example accident black spots, or pedestrian flows or key species if you are doing a lichen survey. You can also take photographs of key landscape features or soil profiles. Figure 1.20 shows a sample quadrat and has been annotated to show marker species and per cent cover.

5. Photographs are absolutely vital in the method section if you have developed a visual index or a quality index. Use a set of photographs as a scale of visual quality to show how you awarded points for the quality of landscapes, shops, or houses.

6. Photographs can be used as part of an impact survey. Take a photograph of the landscape or townscape as it is now, then superimpose what it will look like after the site has been developed or the bypass has been built. This can be done with an acetate overlay or you may be able to scan the photograph into a computer or use a digital camera. In Figure 1.22 the student has marked on the proposed site of a new very large gravel quarry to show how it will affect the landscape.

Figure 1.20 A photograph of a sample quadrant with annotations

show marker species

% of various types of species

show % bare ground

Figure 1.21a Zone of discard/decay

Figure 1.21b Zone of regeneration/assimilation

If you can gain access to a digital camera these are **very** useful as you can take a huge range of shots cheaply, choose the best ones and scan the results in straight away.

Figure 1.22 A photograph of the proposed site of a new very large gravel quarry

Extent of quarry (mark on map)

Evaluate loss of farmland and ecosystems

review any hydrological implications

Assess potential visual impact and landscape quality loss

RIVERS AND VALLEYS

RIVER CHANNEL VARIABLES: BASIC SURVEY TECHNIQUES
River cross-sections
River velocity
Internal and external friction
River gradient
River discharge
Sediment analysis

SITE SELECTION
Site selection for studying changes downstream
Detailed studies of a shorter stretch of a larger stream

RIVER WATER QUALITY
Selecting a site
Environmental monitoring
Chemical monitoring

RIVER AND CATCHMENT MANAGEMENT
How and why river corridors vary
Causes and response of flooding
The use (and misuse) of a local reservoir facility

RIVER VALLEYS AND SLOPE SURVEYS
Looking at slopes

HYDROLOGICAL PROJECTS
Secondary data sources and websites

COASTS
The coastal environment

EROSIONAL STUDIES
Cliff surveys
Wave surveys
Historical erosion rates
Longer-term sea level change

DEPOSITIONAL STUDIES
Beach profiling
Longshore drift surveys
Studying processes along a stretch of coast

COASTAL MANAGEMENT SURVEYS
Coastal protection surveys
Tourist/amenity surveys
Beach water and quality surveys
Websites

ICE DETECTIVE

ICE PROJECTS

DISTRIBUTION AND ORIENTATION OF FEATURES
Upland environments
Lowland areas
Other 'distribution' type studies

MORPHOLOGICAL MAPPING

SEDIMENT ANALYSIS
Sediment size
Sediment shape
Sediment orientation
Degree of sediment sorting
How to interpret sediment results

WEATHERING STUDIES

MOUNTAIN PROJECTS
Tourism
Economy and tradition
Physical projects

ACTIVITY AND AMENITY SURVEYS
Websites

WEATHER

WEATHER PROJECTS
Temperature

2 PHYSICAL ENVIRONMENTS

Precipitation
Pressure
Relative humidity
Wind speed and direction
Cloud cover and type
Visibility
Radiation and sunshine

WEATHER STATION STUDIES

MICROCLIMATE STUDIES
Sampling procedures
Topoclimates
School, garden and urban microclimates
Woodland microclimates
Water microclimates
Secondary data sources and websites

ECOSYSTEMS

INVESTIGATING SOILS
Soil survey techniques

Soil assessment in the field
Soil texture
Laboratory-based soil analysis
Soil erosion risk

INVESTIGATING VEGETATION
Vegetation survey techniques
Recording vegetation types
How many samples?

COMMON BRITISH ECOSYSTEMS
Sand dune ecosystems
Hydroseres
Salt marshes
Upland ecosystems
Woodlands
Hedgerows
Gardens

ASSESSING THE HUMAN IMPACT ON ECOSYSTEMS
Basic survey techniques
Websites

Physical environments

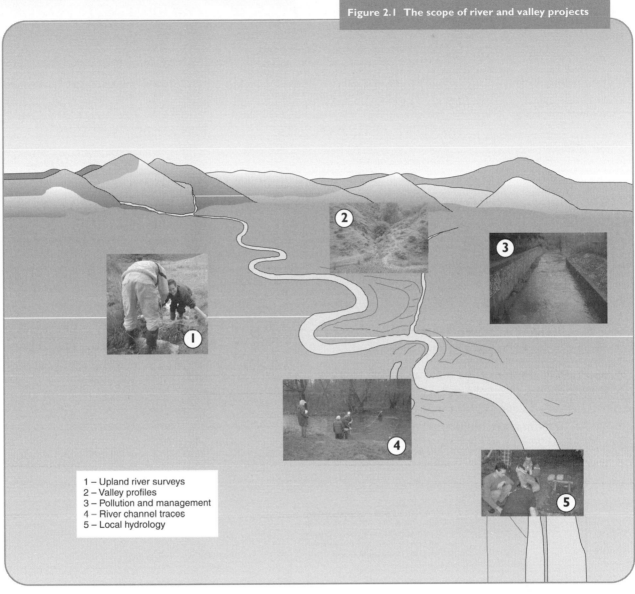

Figure 2.1 The scope of river and valley projects

1 – Upland river surveys
2 – Valley profiles
3 – Pollution and management
4 – River channel traces
5 – Local hydrology

The movement of the Earth's waters from one place to another (and the dissolved loads carried by them) are part of a continuous overall pattern: the hydrological cycle. Rivers and streams are an important component of this cycle, draining water from the landscape and transporting it to the sea. Figure 2.1 shows the range of projects involving rivers. The data collected will often be numeric in format, which makes analysis and interpretation of results more logical and rewarding.

Whilst not all schools or students will have good access to river sites within the immediate locality, this section of the book also suggests ways in which other elements of the hydrological cycle and river catchment can be explored, for example storm hydrographs in the school grounds, water quality issues and river corridor surveys.

Table 2.1 gives some idea as to the scope of river projects, but is not intended to be exhaustive. Group projects are especially well suited to rivers as they allow the collection of a large set of primary data that can be pooled. This means that all group members can select their own topic, based on their own observations, but supported by group (secondary) data.

The range of themes covered in this chapter | Table 2.1

Major topic	Detail	Location
River channel variables	Changes upstream. Variety of variables can be measured.	Upland and valley area.
	Detailed study of interrelationships, e.g. comparison of pools and riffles.	Lowland site.
River channel traces	Studies of sinuosity and meander profiles. Detailed meander studies.	Lowland or upland, but lowland is preferable for most types of study.
River water quality	Use of invertebrates, visual and chemical tests to determine water quality.	Most likely lowland where point sources of pollution can be identified.
River management	Past maps, flooding issues and possibly river corridor surveys.	Most likely lowland, especially when flowing across populated areas.
River valleys	Valley cross profiles and valley gradients, may include sediment analysis.	Usually upland locations.
Hydrological projects	Infiltration on different surfaces.	Can be anywhere – suited to local/school.
	Storm simulation models for a variety of surfaces.	Local, e.g. within school grounds.

RIVER CHANNEL VARIABLES: BASIC SURVEY TECHNIQUES

Figure 2.2 reveals some of the measurements that can be taken at each section of a river channel and a brief description how to carry them out. Depending on the number of recordings taken, you will probably need to allow between 20 and 40 minutes at each location.

TOP TIPS
Field sketches and photos

Field sketches, photographs and site descriptions should also be made at each station, keeping an accurate record of which photograph/sketch etc. is relevant to which site.

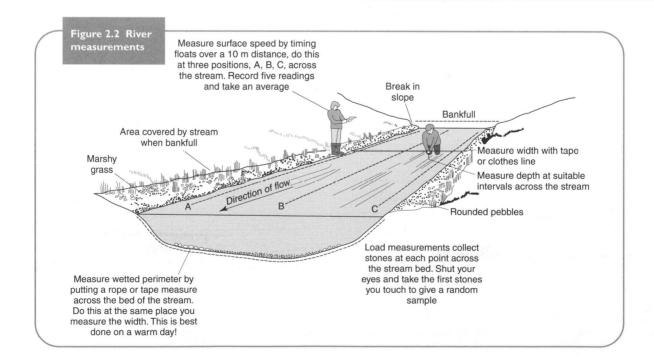

Figure 2.2 River measurements

Measure surface speed by timing floats over a 10 m distance, do this at three positions, A, B, C, across the stream. Record five readings and take an average

Break in slope

Bankfull

Area covered by stream when bankfull

Marshy grass

Direction of flow

Measure width with tape or clothes line

Measure depth at suitable intervals across the stream

Rounded pebbles

A   B   C

Load measurements collect stones at each point across the stream bed. Shut your eyes and take the first stones you touch to give a random sample

Measure wetted perimeter by putting a rope or tape measure across the bed of the stream. Do this at the same place you measure the width. This is best done on a warm day!

River cross-sections

Producing a cross-section of a river channel is a basic fieldwork skill. Whether you need to find out discharge, or examine the profile of a feature such as a meander or riffle, it will be necessary to produce a cross-section of a river. The first stage is to measure the width and depth of the river. The data gathered can then be plotted to create a scale diagram of the cross-section, or used to calculate the cross-sectional area and wetted perimeter of the river.

Channel width (m) Use a tape measure secured at the zero end and measure the width across the surface of the stream at 90° to the course of the river. Try to avoid drag of the tape on water surface by holding it a few centimetres above the water. It is possible that the river or stream only occupies part of the whole channel (e.g. it is a braided section). If you wish to include a prediction of the river's characteristics when in flood, you will also need to measure the bankfull width. If the stream width is very variable then consider taking an average for your particular reach or section.

Bankfull width (m) A tape measure should be stretched across the river from one bank to the other (at 90° to the course of the river). The start and finishing points for the tape are the places where the vegetation and gradient of the bank suggest the river has reached its maximum height.

Channel depth (m) This should be taken at regular intervals across the river, dependent on its stream width. Aim for about 6–8 measurements, and use a metre stick or equivalent to record depths as well as noting down the sampling interval which should be regular, e.g. 20 or 30cm.

River velocity

The velocity of a river is the speed at which water flows along it. The velocity will change along the course of any river, and is determined by factors such as gradient, volume, the shape of the river channel and the amount of friction created by the bed, rocks and plants. Various methods can be used to measure velocity.

Flow meters (m/s, msec^{-1}) These are at the more sophisticated end of the market, such as a 'hydroprop' or similar (Figure 2.3), however they are not cheap at around £100 each. Face the impellor (screw device) upstream at the same points as the depth intervals and calculate a mean flow rate. This will involve conversion of the raw count-rate data

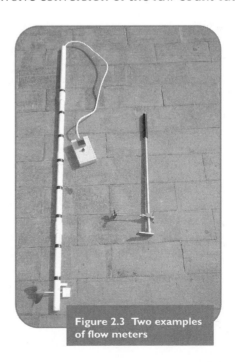

Figure 2.3 Two examples of flow meters

into a useful flow velocity unit. Velocities can also be recorded at different depths.

Floats (m/s, msec⁻¹) Alternatively, you can use a float such as a dog biscuit or orange to measure the *surface speed* of the stream. Use a float which is highly visible, and something which does not catch the wind. Best results will be achieved with timings made over a 5–10m stretch of the river, with repeat sampling to obtain a reliable mean. Speed can be calculated by:

$$\text{Speed / Velocity} = \frac{\text{Distance (m)}}{\text{Time (secs)}}$$

e.g. if a float travels 10m in 21 seconds, its velocity = 10/21 = 0.48 m/sec

TOP TIPS — The 'Float Fudge Factor'

You can multiply your average velocity by 0.85 which is a correction factor. The float tends to follow the line of fastest flow in the river as it is on the surface. This 'fudge factor' reduces the velocity measurement to a more reliable mean.

Salt dilution gauging (m/s, msec⁻¹) This method can also be used to approximate flow velocity, providing you can get your hands on the necessary equipment. It is based on the concentration of solutes in the stream which carry an electrical charge and act rather like an invisible dye. A conductivity meter will measure the water's ability to transfer this charge.

Measure out a 10 or 20m stretch of river to use for the reach. Then dissolve a handful of salt in a bucketful of water and empty this quickly into the upstream section of the reach. Start timing immediately. At the bottom of the section record the conductivity every 3 seconds; the concentration of salt will pass as a wave when plotted against time (Figure 2.4). Observe the time (in seconds) to peak concentration and divide this by the distance travelled.

As in other velocity measurements, it is good practice to repeat the process to obtain a reliable mean.

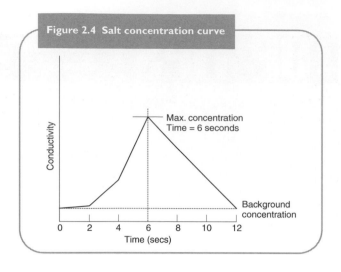

Figure 2.4 Salt concentration curve

Internal and external friction

Water is often considered to be a lubricant, so the idea of friction may not always be obvious. However, when two touching items move past each other there will be friction. In a river the most obvious moving item is water – as water flows over the bed and banks of a river it experiences friction in the form of drag. At the water surface there is also friction generated by air moving across the water.

Estimating friction The frictional forces acting in a river can be divided into two types, internal and external, based on where friction occurs. You might find it useful to use a friction table (Table 2.2) when making these measurements in the field.

Internal friction is also known as turbulence. It can be visually estimated by deciding how calm or rough the flow seems. Very calm indicates limited friction, whilst fast flowing white water has higher friction. External friction is the water touching the bed and banks. The amount of friction depends on the bed, banks and shape of the channel. Smooth channels will have low drag, whilst rough channels will have a higher friction. Use Table 2.2 to award friction points to different parts of the river where samples are taken.

Wetted perimeter (m) This is the total length of a river channel in cross-section which is in contact with the water. Use a tape measure, thick rope or chain, starting at the water level on one bank. Work across the river bed trying to keep the tape (rope or chain) in contact with the bedload at all times (Figure 2.5). This is difficult to measure accurately in the field, particularly if the river is fast flowing and the bed is rough.

	Friction points	Internal friction – turbulence	External friction – banks and bed	
Low Friction	1	Very calm water. Associated with lowland rivers.	Smooth silty bed, no weeds or roots. Uniform channel profile.	
	2	Water moving in the channel at different speeds.	Fairly smooth bottom, sandy and free of weeds.	
	3	Some white water in parts of the channel.	Undulating bottom with some vegetation, sand and gravel on the bed.	
	4	White water moving over large boulders in most parts of the stream. Typical of upland streams.	Irregular bottom with weeds and coarse gravel subrate.	
High Friction	5	Very severe white water and rapids. Found in steep upland and mountainous areas.	Very irregular bottom with large boulders, steps and many weeds.	

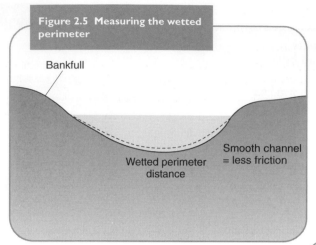

Figure 2.5 Measuring the wetted perimeter

Bankfull

Wetted perimeter distance

Smooth channel = less friction

TOP TIPS

Other ways of deriving of wetted perimeter
It is equally possible to approximate the wetted perimeter from a scaled cross-section drawing, or to calculate it from simply adding depth ($\times$ 2 for each side) to the width to give linear distance.

Hydraulic radius (no units) This is a measure of the efficiency of a channel and is expressed as a ratio. Figure 2.6 shows two contrasting stream forms, where Stream B is more efficient than Stream A as there is more water in relation to the wetted perimeter and therefore less friction.

$$\text{Hydraulic radius} = \frac{\text{Cross sectional area}}{\text{Wetted perimeter}}$$

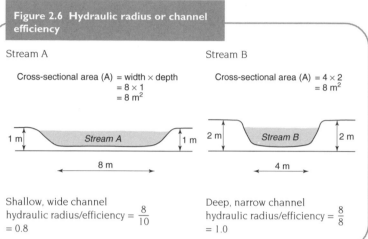

Figure 2.6 Hydraulic radius or channel efficiency

Stream A

Cross-sectional area (A) = width $\times$ depth
= 8 $\times$ 1
= 8 m^2

1 m   *Stream A*   1 m

8 m

Shallow, wide channel
hydraulic radius/efficiency = $\frac{8}{10}$
= 0.8

Stream B

Cross-sectional area (A) = 4 $\times$ 2
= 8 m^2

2 m   *Stream B*   2 m

4 m

Deep, narrow channel
hydraulic radius/efficiency = $\frac{8}{8}$
= 1.0

River gradient

The gradient of a river is a measure of how steeply it loses height.

Using a basic clinometer (°) You need to measure short stretches of the river, where your particular reach is located. The length that is measured should be decided on site, depending on variation in the river itself and the degree of detail required, but somewhere in the region of 10–30m should be ideal. Using two ranging poles or metre sticks, locate these at either end of the stretch, in the middle of the river. Place the clinometer at a comfortable height on the pole and point the clinometer at exactly the same height on the other pole (Figure 2.7). Then record the value.

Using the 'abney level' or 'sunto' style clinometer (°) If you are lucky enough to get hold of one of these clinometers, then these are preferable to the cheaper plastic versions. Use them in the same way described above, but they should give more reliable and accurate results.

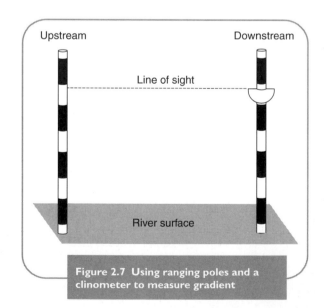

Upstream   Downstream

Line of sight

River surface

Figure 2.7 Using ranging poles and a clinometer to measure gradient

River discharge

The discharge of a river is the volume of water which flows through a measured cross-section in a given amount of time. Discharge varies both spatially and temporally, and is typically controlled by such factors as climate, vegetation, soil type, topography and relief, and the activities of people.

Calculation of discharge (m³/sec) Discharge is very easy to calculate, being the cross sectional area of the channel, multiplied by the velocity of the water.

Calculation of cross sectional area There are two ways of doing this. The most basic is to multiply the width and depth together, as in Figure 2.6. A more preferable method is to plot the cross-section on to graph paper, as in Figure 2.8. Once this has been drawn, it is possible to find the area by simply counting the number of squares in the 'wet' part of the diagram. This method works well for small streams, but the number of squares on the graph paper rapidly goes up for larger streams.

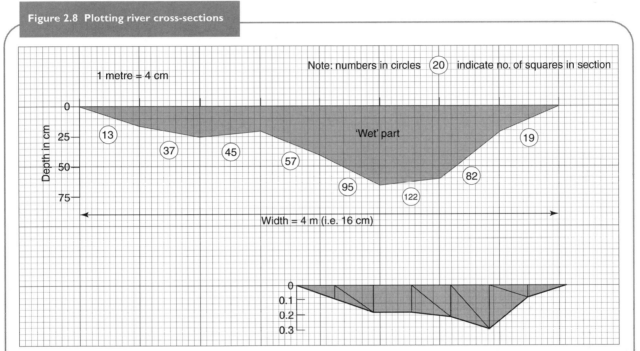

Figure 2.8 Plotting river cross-sections

Area = 470 squares. This number has to be converted, according to our scale. The scale 4 cm = 1 m means that each millimetre square on the graph represents 0.0025 m² in reality (¹/₄₀₀). If the scale had been 1 cm = 1 m then the scaling factor would be ¹/₁₀₀ = 0.01 m². In this example 470 × 0.0025 = 1.175 m²

TOP TIPS

Triangular cross-sections
An alternative to the simple 'counting squares' method or basic cross-section methods is to break up the cross-section into triangles and rectangles, finding the area of each and adding them together.

Calculation of average velocity Remember that the velocity within a river channel varies according to depth and nearness to obstacles such as the bed, banks and boulders. If you have just recorded the surface velocity, then scale this down using the 'fudge factor' (see page 30). Make sure the velocity is converted from a time measurement (i.e. in seconds) to a measurement in m/sec.

TOP TIPS

Incorrect discharge calculations are made when students do not use the correct units, i.e. centimetres are used instead of metres. For most streams you will work on you should expect a low discharge figure, *in many cases much less than 1m³/second.* Even

Britain's biggest rivers such as the Thames and Severn only have average discharges which are in the order of 20–50 cumecs. Bear in mind, that each cumec is equivalent to 1 tonne of water travelling at 1 metre per second.

Load analysis

Rivers carry three types of load – bedload, suspended load and solution load. Load is important in a river as its size and shape can help tell us something about the processes that are occurring within a particular stretch or reach. Figure 2.9 shows how you might expect bedload size to change downstream. Further details for size and shape can be found in the sediments section p. 80.

   Assessing bedload. Two main methods can be used to survey bedload within a stream.

1. Take samples of boulders and pebbles from the stream bed and see how the results vary within different stations or reaches. It is best to use a random method to collect the bedload for a particular reach. Insert a metre stick at intervals across the width of the river bed and collect stones which are touching the stick. Aim for a minimum of 10 stones from each reach (taken from across the stream), although a larger sample of 20–30 stones is preferable. An alternative technique is to collect a small bucket of stones from the bed and weigh this back at school. Using the same technique at each reach will ensure some degree of fair testing. Once the sample is obtained, measure the long axis, shape and radius of curvature of each pebble and then derive the index of roundness.

2. To assess stream competence, collect 20 well rounded pebbles for each of the four size

Group A	1.0–2.9 cm
Group B	3.0–5.9 cm
Group C	6.0–8.9 cm
Group D	>9.0 cm

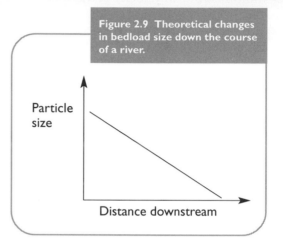

Figure 2.9 Theoretical changes in bedload size down the course of a river.

categories suggested here. Note sizes are based on lengths of long axes.

   Colour-code the pebbles according to category, using a high visibility paint and then line them up at a marked point across the stream. Record the travel paths of the stones over time, plotting distances and any lateral movements within the channel on to a prepared base map. A flow line diagram might be suitable for this purpose (Figure 2.10). It will take a few days or even weeks for some of the larger stones to be moved, so this type of technique is only suitable if you are able to go back and revisit your stream. Try to relate changes in river velocity or discharge to distances moved. Although large stones are usually only moved in storm events.

   Assessing **suspended load.** This is composed of finer material, e.g. sand, silt or clay, which is entrained in the river flow. The movement of suspended load might be related to discharge.

   At the location being studied, secure a bottle to the stream bed with the opening facing upstream and wait for the bottle to become filled with

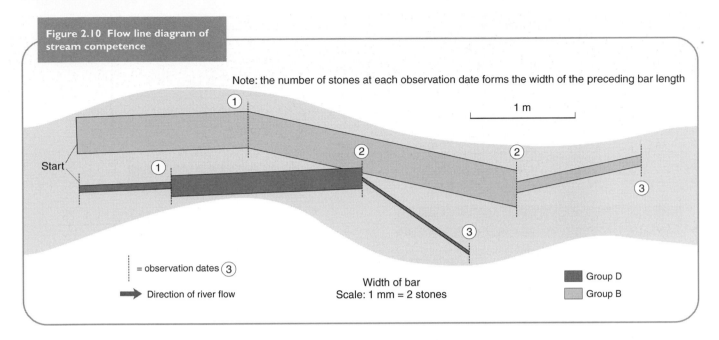

Figure 2.10 Flow line diagram of stream competence

Note: the number of stones at each observation date forms the width of the preceding bar length

1 m

Start

| = observation dates ③
→ Direction of river flow

Width of bar
Scale: 1 mm = 2 stones

■ Group D
■ Group B

---

sediment. To ensure a fair test, make sure the procedure is repeated for the same time and in the same way at any other sites being investigated.

Once the sediment has been allowed to settle, measure the sediment thickness and comment on the particle size, i.e. small, medium, large.

It is possible to calculate the suspended sediment concentration for the river. Filter the mixture of

sediment and water, and weigh the total amount of oven dried sediment remaining on a piece of fine filter paper. Convert this figure to the amount of sediment (grams) per litre. This is easy if your plastic bottle is 1 litre exactly, otherwise scale up or down accordingly. If the discharge of the site is also known (measured in cumecs), then it is possible to calculate the amount of sediment passing per second:

Suspended sediment concentration = 5.2 grams per litre (calculated from the oven drying)

Discharge = 0.021 cumecs or 21 litres / second (litres per second)

Therefore, 5.2 x 21 = 109.2g sediment passing per second.

Assessing solution load. This was originally rock material (i.e. chalk and limestone) which has been dissolved and is now carried within the water of the stream.

A conductivity meter is needed to measure solution concentrations. Using a repeat sampling procedure for the selected reach, record the conductivity of the water at various locations, until a reliable mean or average can be obtained. The higher the reading, the greater the solution concentration, but this will vary considerably depending on the underlying geology. It may be possible to investigate changes in the solution load before and after a storm event, or at varying locations along the course of a river.

SITE SELECTION

Figure 2.11 shows some of the considerations to bear in mind when choosing a stream to study. Once you have decided on your stream, you will have to decide what type of study you will be carrying out. You should also design any booking or recording sheets that will be required before going out and collecting the data. It is a good idea to copy more sheets than you think will be needed so that there are spare ones in case any get spoilt. Figure 2.12 is an example booking sheet for river measurements.

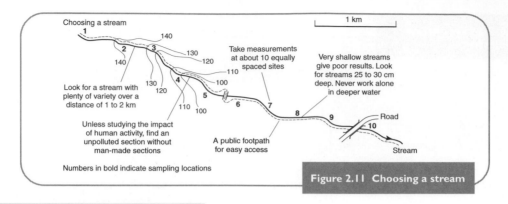

Figure 2.11 Choosing a stream

Figure 2.12 Recording sheet for river measurements

RIVER STUDIES RECORDING SHEET

Site / Reach number..........   Grid Reference.........   Date..........

Measuring the cross-section

Width of stream........(m)   Bankfull width............(m)
Depth of stream at equally spaced intervals. Interval.................
1.................   2.....................   3..................   4..................   5..................
6.................   7.....................   8..................   9..................   10..................
Extra depth at bankfull............(m)

Measuring the speed or velocity

Time taken for float to travel a distance of............(m)
1.........(sec)   2.........(sec)   3.........(sec)   4.........(sec)   5.........(sec)
6.........(sec)   7.........(sec)   8.........(sec)   9.........(sec)   10.......(sec)

Gradient, friction and wetted perimeter

Slope angle...........(°)   Visual score for external and internal friction:............and............(points)

Wetted perimeter..........(m)

Calculations

Average depth............(m)   Hydraulic radius............(points)
Cross-sectional area..............(m²)   Average velocity...........(m/sec)
                    Discharge...............(cumecs)

Site selection for studying changes downstream

▶ Using 1:50,000 and 1:25,000 scale maps, decide on the channel length that will be surveyed and how many sampling points will be used. The ideal number is somewhere between 8 and 12 stations. This will depend on the number of helpers, daylight time and the overall distance that will be walked.

▶ Decide on the sampling strategy. Most river studies use either a systematic or stratified framework. Take account of access and obvious sampling locations, e.g. bridges or footpaths and think about repeat sampling, either at each station or on different days. Identify likely points for sampling (which may be separated by a matter of meters or a number of kilometres) and map them onto a base map. Remember that you will need at least eight sites to perform statistical analysis, e.g. Spearmans Rank. Figure 2.13 shows an example data set and scatter graphs.

Figure 2.13 Downstream changes in hydraulic variables

	Distance downstream from site 1 (km)	Width (m)	Mean water depth (m)	Wetted perimeter (m)	Mean flow velocity (m/sec)	Discharge (cumecs)	Amount of load in suspension (g/l)
Site 1	- - -	1.7	0.24	2.58	0.44	0.18	0.37
Site 2	2.0	0.96	0.26	1.48	0.46	0.11	0.23
Site 3	2.2	1.60	0.29	2.30	0.04	0.02	0.50
Site 4	3.8	2.30	0.32	3.10	0.23	0.17	0.67
Site 5	4.1	2.10	0.46	3.00	0.42	0.41	0.62
Site 6	4.9	3.90	0.57	5.20	0.64	1.42	0.95
Site 7	6.1	3.30	0.44	4.25	0.61	0.89	0.15
Site 8	6.4	3.70	0.46	4.65	0.73	1.24	1.12
Site 9	7.8	4.10	0.49	5.10	0.70	1.41	0.98
Site 10	8.2	3.80	0.58	5.15	0.66	1.45	0.85
Site 11	9.0	2.00	0.41	2.63	0.58	0.48	0.44
Site 12	11.0	3.90	0.43	4.80	0.46	0.77	1.32

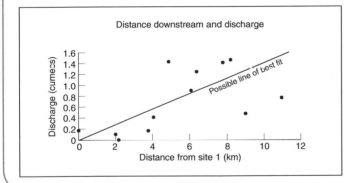

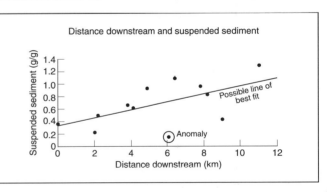

TOP TIPS

Your project write up should include a justification for both the sampling framework and individual stations chosen. Even if you go on a field course and feel that these decisions are being made by other people such as your teacher, tutor or other students, it is wise to question why and how particular sites or reaches have been chosen.

▶ Carry out a pilot study to verify access to sampling stations and get written permission from landowners if you are using any stretches of private land. You should also carry out a risk assessment at this point.

▶ It is perhaps more logical to begin recording upstream and work down towards the source, but for group investigations, your teacher/tutor may decide that the surveys are to be carried out in the other direction because of transport logistics. This is not a problem, but make sure you know how the sites are being selected and in what order you will be expected to work.

Examples of projects include the following.

Title/Aims/Hypotheses	Details

1. An analysis of stream channel characteristics to see how they vary downstream, i.e. does the stream get wider, deeper, flow faster and become more efficient downstream?
 	
Example hypotheses:
 	
▶ There is a proportional increase in discharge with increasing distance from the source.
 	
▶ There is a significant, positive, correlation between width and depth.
 	
▶ There is a negative relationship between gradient and velocity.
	This is the classic river study, and can work equally well on either a group or individual basis. Use the survey methods outlined above, but if you are working on an individual basis, do not attempt everything. Try and keep the project focused and manageable. In a group situation it is likely that you will collect some data that contributes to the whole data set. Again, be selective in what you use and keep it workable.
	   Tabulate your results and use Spearman's Rank to validate any hypotheses.
	   Try to support your project with secondary data. If you are lucky you may be able to use past student data, or even data from the Environment Agency.

2. How does the load vary in size and shape along the stream's course? Does the sediment fit the predicted theory, i.e. become smoother and smaller downstream? Example hypotheses: ▶ There is a proportional decrease in sediment size with progression downstream. ▶ There is a tendency for sediment to become more rounded with distance away from the source.	Select 10–12 sites along a stretch of river, preferably choosing locations where there is likely to be a change in sediment, i.e. just below confluences (use a map to assist with this). Ensure you have access to the sites and that they are safe. Using a stone board collect sediment samples from each reach, perhaps 20 at each site. Obtain a mean long axis measurement and shape description (either Power's or Cailleux, see page 79) for each site. It is also a good idea to calculate the gradient between each site. Data can be plotted as histograms along a sketched/plotted long profile. Use Spearman's Rank to test for relationships.
3. Is there a testable relationship between hydraulic radius and river velocity with progression from source to mouth? Example hypotheses: ▶ There is a significant, positive, correlation between velocity and hydraulic radius. ▶ There is no significant relationship between velocity and hydraulic radius.	Use a similar sampling strategy to the one recommended in part (1). Measure velocity using two methods at each site, i.e. float and flow meter, and calculate a reliable mean for each reach. Survey the cross sectional area and measure the wetted perimeter to obtain the hydraulic radius at each station. Use Spearman's Rank to obtain a correlation coefficient and plot scatter graphs. Try to suggest reasons for any changes you have identified, i.e. changing geology or channel morphometery.

TOP TIPS — Null and alternative hypotheses

When using a test such as Spearman's Rank, it is good practice to develop null hypotheses for each alternative prediction/hypothesis that is made, i.e.

H_1(alternative): There is a proportional increase in discharge with increasing distance from the source.

H_0 (null): There is no relationship between the volume of discharge and distance downstream

The reason for the so-called 'null' hypothesis, is because the statistical test actually looks at the validity of this response and then either rejects or accepts it. In other words, if the null hypothesis is rejected, the alternative hypothesis is accepted.

Detailed studies of a shorter stretch of a larger stream

Many larger rivers are too big and too dangerous to survey, so much of the fieldwork for this type of survey should be well supported with maps and secondary data. Some large lowland streams, however, are safe to work in, especially during the summer months when the water level is at a minimum. Your teacher will be able to help and advise you on safety issues. Below are some project ideas that can be investigated.

Title/Aims	Details
1. A detailed survey of depth and speed variations within a meander and straight stretch.	The fieldwork for this is based on a detailed survey of two contrasting sections of river. Start by collecting detailed velocity measurements in the straight section, using a flow meter. This can either be surface speed measurements, or more preferably, within a river cross-section. Repeat the procedure for a straight section. You will need a minimum of 15 points, but ideally aim for 25–30 measurements in the cross-section.
2. A study of the changing characteristics of a river's load between pools and riffles.	Use the approaches outlined on page 34 to investigate the river's load.
3. Investigating an assumed constant section of river (i.e. no inputs) and examining the reliability of velocity or other measurements.	Remember that there are various ways to measure velocity: float, flow meter and salt dilution gauging. Use these techniques at a variety of locations and calculate their reliability – which techniques produce results with the lowest standard deviation? Does one particular method show results which are significantly different to the other two? You might also look at the variation in discharge measurements along different stretches and determine which method of calculation is the most accurate. If possible, use secondary data to support your findings.
4. The relationship of a meander belt to meander models: shape, sinuosity and frequency.	This is a classic lowland river study, being very much based on secondary map evidence. From a large scale map it is possible to determine: flood plain width, river width, wavelength, amplitude etc. Use the models available in textbooks to see whether your river fits the trend. If not, why not? Approaches to this type of investigation are outlined in the next section.

River traces – sinuosity and meander profiles

It has already been suggested that many lowland rivers are simply too deep and too dangerous to survey in their entirety. However, it is likely that access points can be found or negotiated, e.g. footbridges, jetties, so that approximate readings of width and depth can be attempted for a limited stretch of the river. It is feasible to use a 1:10,000 or larger scale map to obtain some of the basic river dimensions.

For each river bend calculate sinuosity (see the box opposite). You will probably need to survey 8–10 identified meanders (with an index of sinuosity over 1.5), but remember this can be achieved using a group approach. Typically the following measurements are made at each point: width (this could include flood plain and/or bankfull), depth and velocity (Figure 2.14).

Sinuosity

For river bends in large, lowland rivers, you can work out an **index of sinuosity**, which relates to the 'bendiness' of a meander or particular stretch of river:

$$\text{Index of sinuosity} = \frac{\text{Distance AB along river}}{\text{Distance AB in a straight line}}$$

Sinuousity is normally in the range 1.5 – 4

TOP TIPS

Remember to support your project with a detailed sketch map(s) and photographs wherever possible. This may mean taking photographs and making sketches of river channel features, but also photographs and sketches of people collecting data, and even the equipment itself.

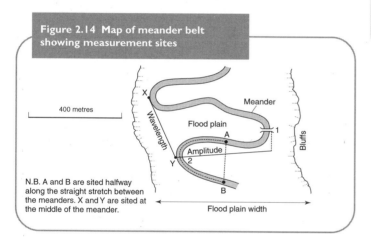

Figure 2.14 Map of meander belt showing measurement sites

400 metres

Meander
Flood plain
A
1
Amplitude
2
Wavelength
Bluffs
X
Y
B
Flood plain width

N.B. A and B are sited halfway along the straight stretch between the meanders. X and Y are sited at the middle of the meander.

Detailed meander surveys

In addition to the basic meander surveys outlined above, it is possible to carry out a number of small scale, more detailed surveys that could form the basis of a whole investigation in their own right (Figure 2.15).

▶ **Meander maps** Figure 2.16 shows a detailed (isoline) depth map for a meander. In this instance over 50 depth measurements have been taken using a metre stick (they have been taken in a regular grid pattern). The results are then plotted onto paper and a depth map constructed.

▶ **Meander speed profiles (isovels)** Take a number of detailed depth measurements and plot a cross sectional diagram. At the same spot, using the

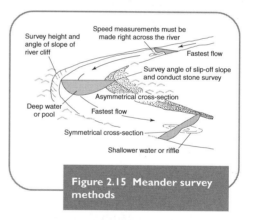

Survey height and angle of slope of river cliff
Speed measurements must be made right across the river
Fastest flow
Survey angle of slip-off slope and conduct stone survey
Asymmetrical cross-section
Deep water or pool
Fastest flow
Symmetrical cross-section
Shallower water or riffle

Figure 2.15 Meander survey methods

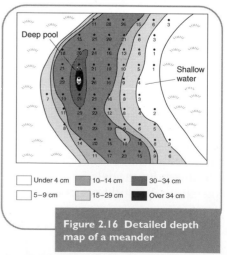

Deep pool
Shallow water

| | Under 4 cm | | 10–14 cm | | 30–34 cm |
| | 5–9 cm | | 15–29 cm | | Over 34 cm |

Figure 2.16 Detailed depth map of a meander

flow meter, record at least 15 depth measurements in a systematic manner to achieve a grid pattern. Plot the results as an isovel diagram (Figure 2.17).

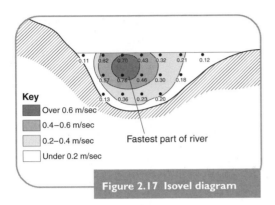

Key

▨	Over 0.6 m/sec
▨	0.4–0.6 m/sec
▨	0.2–0.4 m/sec
☐	Under 0.2 m/sec

Fastest part of river

Figure 2.17 Isovel diagram

▸ **Meander sediment surveys** A number of approaches can be used here. Comparisons can be made of sediment size and shape between contrasting meanders, or a survey of changes in sediment size from the inside to the outside of the bend can be undertaken. In either case aim for a minimum sample of 50 stones at each location. Alternatively studies can be made relating to changes in the nature of sediment on a slip-off slope/point bar. Measure the dimensions of the point bar and then conduct a sediment survey, measuring the size and shape of stones on a mini-transect from the water's edge up to the river bank.

▸ Is the current/flow rate always fastest on the outside of a meander and slowest on the inner bend? Does the line of fastest flow (thalweg) swing from side to side? Suggest reasons for this supposed relationship(s) and why the river may not fit the predicted model.

▸ Is there a relationship between meander wavelength and discharge, or between stream width and wavelength?

▸ What is the relationship of meander amplitude to floodplain width? Does it vary for different (lowland) reaches of the same river, or ever for different rivers?

▸ Is a river's sinuosity always in the range 1.5 – 4? If not, what reasons might put it outside of these ranges? Is wavelength/width normally in the range 7–10?

▸ A study of river cliff and slip-off slope geometry. How and why do these vary for a short stretch of river?

RIVER WATER QUALITY

Rivers are not only essential for supporting freshwater ecosystems, they are also a source of drinking water and sites of recreation and amenity. Rivers are also used as sites for waste disposal, which is regulated in England and Wales by the Environment Agency. Freshwater pollution therefore has immediate social and economic consequences.

Human sewage, animal waste, silage and the byproducts of food processing are all types of *organic* effluents. Bacteria use oxygen to break down these unstable compounds. Nitrate pollution from fertilizers is often considered alongside organic pollutants as it has a similar effect by causing a reduction in water oxygen levels.

Selecting a site

Most studies involve the examination of organic pollution from a single or point source of discharge, e.g. effluent from factories, farm drains or settlements. Choose a stretch of river or other water course (like a canal), consider map information, local issues and any other evidence which may indicate potential sites of pollution. It may be possible to get hold of the 'LEAP' reports from the local Environment Agency offices which detail water qualities for selected rivers. Select an appropriate sampling strategy, usually 'above and below' or 'before and after' the outfall/source of pollution. Ensure that the river section under study is a manageable size for the scale of the study and that there is access to sites which will be under investigation. The number of sites chosen will depend on the nature of the river and the type of study being undertaken. Generally 3–5 sites will be suitable, although it is possible to use just two locations, i.e. above and below the point source. It may be that a group approach is to be used, in which case small groups may be allocated just one stretch of river. Figure 2.18 gives some possible locations for study along the River Perry, North Shropshire.

TOP TIPS

LEAP reports

Your local Environment Agency offices may be able to supply you with a copy of their LEAP (Local Environmental Action Plan) report. Although usually a weighty document, it contains very useful secondary data about the water quality of rivers and streams within the Agency's region.

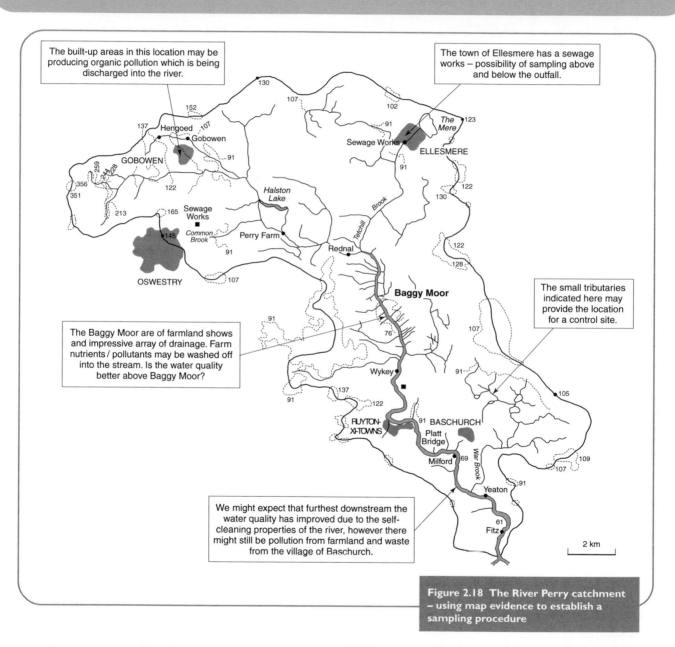

The built-up areas in this location may be producing organic pollution which is being discharged into the river.

The town of Ellesmere has a sewage works – possibility of sampling above and below the outfall.

The Baggy Moor are of farmland shows and impressive array of drainage. Farm nutrients / pollutants may be washed off into the stream. Is the water quality better above Baggy Moor?

The small tributaries indicated here may provide the location for a control site.

We might expect that furthest downstream the water quality has improved due to the self-cleaning properties of the river, however there might still be pollution from farmland and waste from the village of Baschurch.

Figure 2.18 The River Perry catchment – using map evidence to establish a sampling procedure

Two main categories of techniques can be used to determine stream or river water quality: environmental and chemical.

Environmental monitoring

Environmental monitoring relies on a range of approaches:

1. Biological indicators (animals and plants) Some species of freshwater organism have a low tolerance to pollution. The types, numbers and distribution of these indicator species can be used to gauge the quality of water (Figure 2.19). There is usually a progressive loss of clean water animals the closer you get to the source of pollution, and the presence of certain plants is an indication of pollution, e.g. sewage fungus and blanket weed. It is good practice to use biotic indices such as the Trent Biotic Index as a more quantitative method to assess your results (see Top Tips Box).

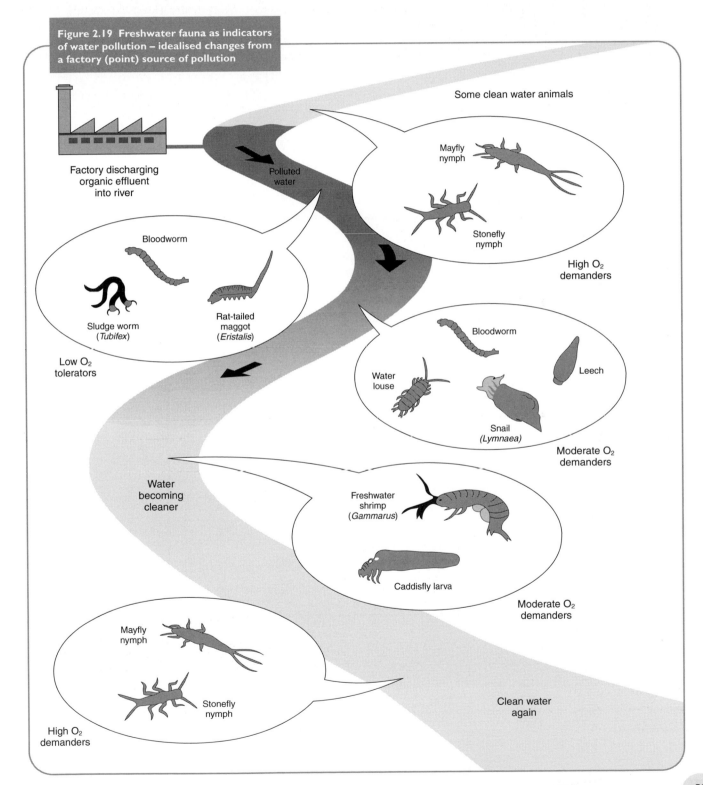

Figure 2.19 Freshwater fauna as indicators of water pollution – idealised changes from a factory (point) source of pollution

Factory discharging organic effluent into river

Polluted water

Some clean water animals

Mayfly nymph

Stonefly nymph

High O_2 demanders

Bloodworm

Sludge worm (*Tubifex*)

Rat-tailed maggot (*Eristalis*)

Low O_2 tolerators

Bloodworm

Water louse

Leech

Snail (*Lymnaea*)

Moderate O_2 demanders

Water becoming cleaner

Freshwater shrimp (*Gammarus*)

Caddisfly larva

Moderate O_2 demanders

Mayfly nymph

Stonefly nymph

High O_2 demanders

Clean water again

TOP TIPS

Biotic Indices

At their simplest level, biotic indices are based on the assumption that there is a progressive loss of clean water fauna with an increasing oxygen demand. Biotic index systems have been developed which give numerical scores to specific "indicator" organisms at a particular taxonomic level. Such organisms have specific requirements in terms of physical and chemical conditions. Changes in presence/absence, numbers, morphology, physiology or behaviour of these organisms can indicate that the physical and/or chemical conditions are outside their preferred limits. Presence of numerous families of highly pollution tolerant organisms and an absence of high oxygen demanders usually indicates poor water quality.

Trent Biotic Index

The first stage in the calculation of the biotic index is to count how many groups of invertebrates are present at each site. The term 'group' in this instance means any one of the sets of organisms included in the table below. For some groups with easily distinguishable species, i.e. the Mayflies, you are asked to identify and count the separate types.

THE GROUPS	THE SITES			
	1	2	3	4
Each known species of Stonefly (Plecoptera)	2			
Each known species of Mayfly (Ephemeroptera)	3			
Each family of Caddis-fly (Tricoptera)	2			
Each known species of hog-lice and shrimps (Crustacea)	2			
Non-red Chronomids	1			
Worms (Annelidae)	0			
Each known species of flatworms (Platyhelminthes)	0			
Each known species of alderfly (Neuroptera)	0			
Each known species of water mites (Hydracarnia)	1			
Black fly larvae (Simulidae)	1			
Each known species of other fly larvae	0			
Each known species of beetles (Coleoptera)	0			
Each known species of snails (Mollusca)	1			
Each known species of leeches (Hirudinea)	1			
TOTAL NUMBER OF GROUPS	14			

Write the number of groups in the boxes, so if two different types/species of stonefly are found then 2 would appear in here.

To calculate the Trent Biotic Index for each site you will also need to use the table on the next page:

TOP TIPS

Biotic Indices *continued*

Start at the top of the species column and work downwards until a line is reached which fits your sample. Then work across the table and stop when the number of groups found at that site fits the correct range. Using the sample data provided, the score will be 9.

			Total number of groups present						
CLEAN			0–1	2–5	6–10	10–15	16–20	21–25	26–30
						BIOTIC INDICES		
Stonefly (Plecoptera) nymphs present	More than one species		0	7	8	9	10	11	12
	One species only		0	6	7	8	9	10	11
Mayfly (Ephemeroptera) nymphs present	More than one species		0	6	7	8	9	10	11
	One species only		0	5	6	7	8	9	10
Caddisfly (Tricoptera) larvae present	More than one species		0	5	6	7	8	9	10
	One species only		4	4	5	6	7	8	9
Freshwater (Gammarus) shrimp present	All above absent		3	4	5	6	7	8	9
Hog-louse (Asellus) present	All above absent		2	3	4	5	6	7	8
Chronomid worms present	All above absent		1	2	3	4	5	6	6
POLLUTED	All of the above absent		0	1	2	0	0	0	0

An alternative to the Trent Biotic Index is to determine the diversity of invertebrates at each site. To do this, simply count up the number of different types of organisms collected. You could then use a simple scoring system, similar to the one outlined here, to derive a diversity score. To make the score useful, you will need to set you own criteria as to how many organisms are required for each category.

5 (diversity high) excellent/good

4 (diversity slightly reduced) fair

3 (diversity significantly reduced) doubtful

2 (diversity low) poor

1 (diversity very low) bad

Invertebrates can be collected or sampled from a stream using a net with a fine mesh size and a flat base. A 'kick sampling' method is used so that the animals are disturbed from the stream bed and washed into the net. Normally a 30 second kick is used at each station to ensure fair testing. Organisms are then sorted, classified and counted according to type and demand for oxygen. Repeat sampling is required for reliable surveying of invertebrates.

2. Temperature and circulation Temperature partly limits stream oxygen levels; the lower the temperature, the more oxygen can be dissolved in the stream. Temperatures can easily be measured with a (waterproof) thermometer and flow rates using a simple float.

3. Visual evidence Qualitative observations can be made such as smell, water clarity or discolouration, surface scum and the presence of rubbish in the

WATER QUALITY REPORT   NOTES

1. Description of colour/turbidity

2. Is there any obvious smell?

3. Are there any signs of vandalism?

4. Is there evidence of birds, animals etc.?

5. Is there any rubbish/debris in the water?

6. Are there any signs of surface scum, froth, or oil in the water or on the banks. Can the source be determined?

7. Is there any blanket weed?

8. Are there any outfalls present?

Figure 2.20 Example visual evidence sheet

Chemical monitoring

Chemical tests include the following.

1. **Dissolved oxygen** As oxygen is used by bacteria to break down organic pollutants, watercourses with low oxygen concentrations may be suffering from pollution – for example, from raw sewage discharges, paper mills or food processing. Fish are high oxygen demanders, and will not be able to survive in streams with low concentrations of dissolved oxygen. Assessment of oxygen levels (ppm) can be established with a meter and probe, or by use of a specialised chemical testing kit. Typically oxygen levels follow an idealised pattern as shown in Figure 2.22.

water or on the bankside (Figure 2.20). Support your written findings with photographs and field sketches wherever possible.

4. **Turbidity** This is a measure of the amount of suspended sediment present in a water sample. High suspended sediment loadings will give the water a turbid or opaque appearance and this may indicate high levels of organic pollutants. To assess turbidity, use either a Secchi disk or painted tile (Figure 2.21a) which is lowered into the water and the depth at which it disappears should be recorded. Alternatively a turbidity tube can be used in shallower locations to estimate turbidity (Figure 2.21b).

2. **Nitrate and ammonia levels** Nitrates are essential for plant growth, but levels of 50 ppm are the recognised limits for potable (safe to drink) water. Nitrates may be sourced from agricultural activities or from sewage and are associated with eutrophication (the nutrient enrichment of water through the accumulation of soluble residues, i.e. nitrates, which encourage algal blooms). Test strips are available for the field assessment of nitrate levels and ammonia levels.

3. **pH** To measure pH use a simple probe or water colourimetric test. Changes in stream pH are complex, and may not necessarily be related to a particular organic source; however acidification of watercourses may be seen as an issue in its own right and so may be worth investigating. The impact of increased acidification is to cause changes in the ecosystem, especially the numbers of primary producers.

Figure 2.21 Secchi disk (a) and (b) turbidity tube

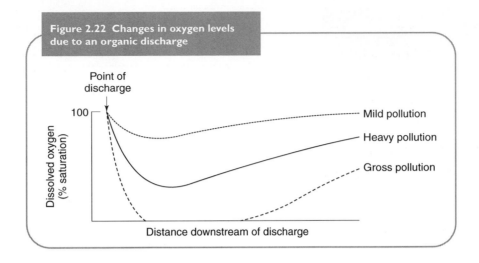

Figure 2.22 Changes in oxygen levels due to an organic discharge

Point of discharge

100

Dissolved oxygen (% saturation)

Mild pollution

Heavy pollution

Gross pollution

Distance downstream of discharge

RIVER AND CATCHMENT MANAGEMENT

The prime task of the water industry in its broadest sense is to manage the hydrological cycle for the benefit of users. Water is extensively used for recreation and amenity and the water industry must not only provide enough water to service all its users, but also supply it to an appropriate standard, together with the management of all its water resources. This is a diverse topic and, there are a number of possible studies you could undertake. What follows are a few examples – you will need to refer to other sections of the text to check some of the techniques, methods and sampling procedures.

How and why river corridors vary?

The river corridor can be defined as the width of river, bank and adjoining land immediately next to a water course. One possible study is to carry out your own river corridor survey. In addition to detailed measurements of the river width etc, you can also map the adjacent land using the symbols shown in Figure 2.23. These types of survey are normally used to identify the wildlife value of particular stretches of river. You might choose to investigate two

contrasting reaches on the same river, or compare different rivers in a variety of locations/land uses. This technique may be useful if you want to create your own management plan for an area.

Causes and responses of flooding

Flooding is a popular topic within AS and A level Geography. Often studies start by providing a base map of the drainage basin in question and then go on to identify the causes and consequences of flooding within particular towns or cities. This type of project can be well supported by secondary data from a range of sources, including the local Environment Agency, local and national newspapers. The primary data collected could be based around a land-use map of a town, identifying areas which are most at risk from flooding. It may also be possible to use the Chi-squared test to investigate the possible relationship between land use on and off the floodplain. An extension of this type of study is to review and evaluate the proposed flood responses, or devise your own management plan. This might be supported with cost-benefit analyses, comparing the costs of building flood defences, versus the potential insurance payout savings (benefits). In order to do this you might have to develop notional land value figures, or get rateable values from the local council.

Figure 2.23 Symbols for use in river corridor surveys

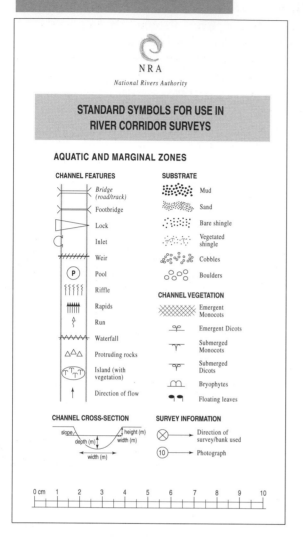

'Benefits' Table of approximated land-use insurance values for each 'square', based on floodplain data and predicted extent of 1 in 100-year flood

Land-use category	No. of squares	Proportion (nearest %)	Cost per unit (£1000's)	Total insurance costs (£1000's)
PRE-WAR LOW COST HOUSING	65	14	42	2730
PRE-WAR HIGH COST HOUSING	57	13	126	7182
POST-WAR LOW COST HOUSING	105	23	42	4410
POST-WAR HIGH COST HOUSING	72	16	126	9072
RETAIL/ADMIN	21	5	164	3444
INDUSTRY	5	1	45	225
TRANSPORT	3	1	22	66
CAR PARKS	19	4	14	266
OPEN SPACE (Including river)	88	20	2	176
AGRICULTURE	12	3	8	96
TOTAL	447	100%	**Total 'insurance' costs = £27.67 million**	

The use (and misuse) of a local reservoir facility

There are many reservoir facilities thoughout Britain, many supplying water for our large towns and cities. These can provide a number of possible studies as outlined below:

▶ **Visitor surveys** Questionnaires can be devised to determine the sphere of influence of particular facilities, and to establish the reasons why visitors may visit such areas. This could also be coupled to an activity survey to establish 'hot-spots' within the reservoir site, i.e. the places that are most popular and the reasons for this.

▶ **Conflicts of interest** Frequently reservoirs can create conflicts of interest amongst a variety of interest groups, e.g. the farming community, conservationists, tourists and water companies, especially through the ways in which the resource is managed or owned. In the Lake District for example, usage of lakes for water-skiing activities has created considerable conflicts and a speed limit has been imposed. Try to gauge a range of people's concerns and develop your own ideas through a management plan.

▶ **The development of a new reservoir** A possible study might be to examine the consequences of a new reservoir, whether real or hypothetical. If hypothetical, use a detailed OS map, and try to select a site which might be suitable for a reservoir (often upland locations are most suitable). Trace the contour lines and approximate the area that might be inundated with different sizes of reservoir within a particular valley. You might then consider evaluating the site within an environmental impact assessment which considers the effects before and after development. Also check the water quality and calculate the stream discharge as further evidence for your report.

RIVER VALLEYS AND SLOPE SURVEYS

The steepness, or gradient, of a slope is one of the main determinants of how the land is used or occupied (see Table 2.3). Generally the steeper a slope the less it can be used for economic activity – some river valleys may exhibit very steep profiles which can only be used for hill sheep grazing or agroforestry.

Land use and steepness of slope | **Table 2.3**

Slope (°)	Typical gradient	Description of steepness	Comment
<1°	----	Level	Very flat.
1° – 3°	1 in 60 (1.7%)	Flat	Ideal for railways and most other uses. Problems of waterlogging may be common.
3° – 6°	1 in 20 (5%)	Gentle	Good for building and agricultural purposes – improved drainage.
6° – 12°	1 in 10 (10%)	Moderate	This feels like a moderate slope when walking up it.
12° – 20°	1 in 3 (33%)	Fairly steep	Limit for mechanised agriculture. A problem for heavy vehicles on roads.
20° – 35°	1 in 2 (50%)	Steep	Roads, tracks and paths will zig-zag to cope with the steep slope.
35° – 45°	1 in 1	Extremely steep	Hands needed to scramble up.

TOP TIPS   Gradient

Slopes are measured in degrees, from 0–90°, whilst the gradient of a slope is expressed either as a percentage or ratio, as on road signs for hills.

To convert the degree of a slope to a gradient use the tangent key on your calculator, e.g. tan 20° = 0.36, i.e. 36%.

Looking at slopes

Slope forms come in all shapes and sizes, although two main types are usually recognised: concave and convex (Figure 2.24). You can use a number of approaches (both primary and secondary techniques) to examine and survey slope patterns.

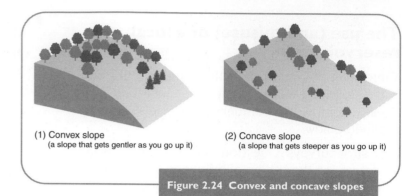

(1) Convex slope
(a slope that gets gentler as you go up it)

(2) Concave slope
(a slope that gets steeper as you go up it)

Figure 2.24 Convex and concave slopes

Gradients	Contour lines on OS maps provide clues as to the nature of slopes. Where they occur tightly packed together the slope angle changes quickly, and where they almost merge the ground could be very steep, e.g. a cliff.
Cross sections	Cross-sections are vertical slices through the land surface. Transect lines on a map can be converted into cross-sections by using the contour lines (Figure 2.25). Use a piece of paper (at least as long as the transect line), place one edge along the transect and mark off each contour intersection, noting the height. Then draw a cross-section as in Figure 2.25, using a suitable scale. Make sure you include the trend line of the transect, e.g. NE to SW.
Aspect	Aspect is the direction in which a slope faces; it may influence vegetation structure and type, drainage and rates of weathering. This can either be determined using map evidence, or in the field with a compass.
Levelling	Levelling is the most accurate, but complex, method of slope surveying. It will only be possible if you have access to surveying equipment, i.e. a 'level' and ranging poles. This type of method is best used for surveying longer stretches of slope, e.g. valley sides or beaches. Figure 2.26 illustrates how to carry your own survey on a beach (along a transect from the sea). The slope is measured as the height difference between two points. The points are identified on ranging poles by sighting through the level. Results can be recorded onto a sheet as in Table 2.4 (page 52). **Note that this method of surveying gives a gradient and not an angle.** You will of course need at least two people to undertake this method.
Pantometer	The pantometer method is easier and quicker than levelling. Two uprights (1m apart) are loosely bolted together with two cross pieces. A large protractor and spirit level are fixed to one upright as in Figure 2.27. Follow a transect line, by stepping the pantometer up or down the slope. This technique is rather time consuming for longer slopes, but should reveal the detail of smaller breaks in slope. It also has the advantage that it can be undertaken by one person.

Figure 2.25 Plotting cross-sections

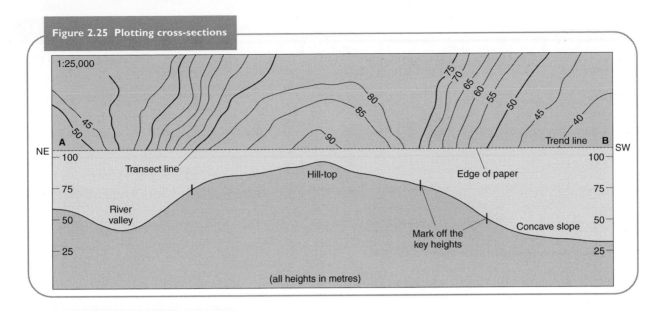

1:25,000

NE — 100 ... SW

Trend line **B**

A

Transect line

River valley

Hill-top

Edge of paper

Mark off the key heights

Concave slope

(all heights in metres)

Figure 2.26 The levelling method

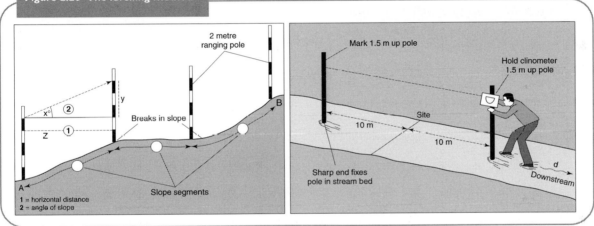

2 metre ranging pole

x° ②

z ①

y

B

Breaks in slope

Slope segments

A

1 = horizontal distance
2 = angle of slope

Mark 1.5 m up pole

Hold clinometer 1.5 m up pole

Site

10 m

10 m

Sharp end fixes pole in stream bed

Downstream

d

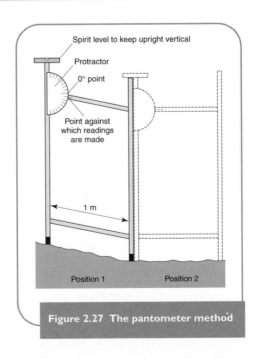

Spirit level to keep upright vertical

Protractor

0° point

Point against which readings are made

1 m

Position 1

Position 2

Figure 2.27 The pantometer method

	Location/ level no.	Backsight height (m)	Foresight height (m)	Difference (m)	Height above sea	Distance between poles (m)	Gradient (°)	Slope (%)
A	1	3.8	2.1	1.7	10	20	4.9	8.5
B								
C								

An levelling recording sheet — Table 2.4

TOP TIPS — Don't overestimate!

The gradient of many slopes may be too shallow or too small to measure using the techniques identified above. In which case, try to estimate the gradient (use any map evidence available to help you). Most people wildly overestimate, by at least a factor of two. Remember that most slopes are much less steep than you might imagine.

There are a range of situations when slope profiling or the determination of slope characteristics might be useful with a piece of fieldwork. Figure 2.28 outlines some of the range of options available.

HYDROLOGICAL PROJECTS

Hydrology is the study of water: where it comes from and how it is used and stored within the environment. Central to the study of hydrology is an

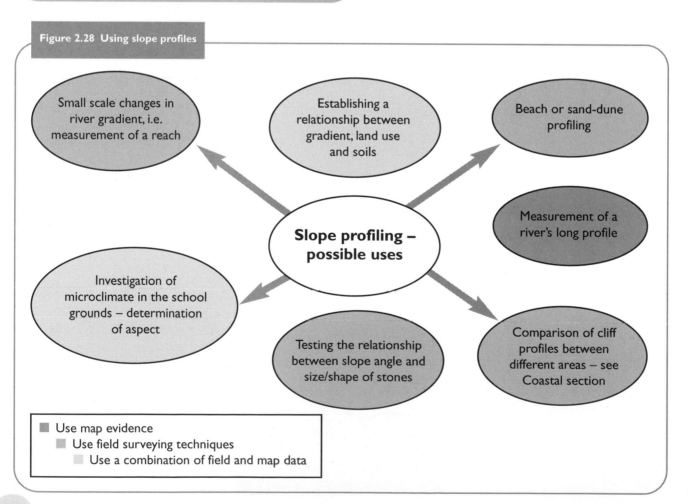

Figure 2.28 Using slope profiles

- Small scale changes in river gradient, i.e. measurement of a reach
- Establishing a relationship between gradient, land use and soils
- Beach or sand-dune profiling
- Measurement of a river's long profile
- **Slope profiling – possible uses**
- Investigation of microclimate in the school grounds – determination of aspect
- Testing the relationship between slope angle and size/shape of stones
- Comparison of cliff profiles between different areas – see Coastal section

- Use map evidence
- Use field surveying techniques
- Use a combination of field and map data

appreciation of the hydrological cycle (Figure 2.29). Most successful projects in this area are limited in terms of scale and centred on the examination of one or two aspects (usually inputs or outputs) of the system.

Hydrological studies are many and varied, and they can be carried out in a range of environments, either urban or rural, or at home or in the school/college grounds. Often these types of investigations require the construction of homemade apparatus or equipment. Many of the experiments can be linked with microclimate and use the same equipment, e.g. rain gauges. A number of possible studies are discussed below.

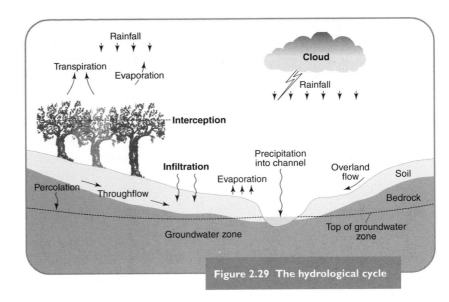

Figure 2.29 The hydrological cycle

Possible study	Data collection and presentation ideas
Storm simulation ▶ To compare the total runoff for different storm surfaces. ▶ To investigate the effects of antecedent (prior) rainfall conditions on successive storm runoff. ▶ To determine how and why the total discharge, peak flow and initial response time vary for different surfaces and storms.	This is an experiment for simulating rainfall events over different surfaces and surface covers, e.g. bare soil, compacted footpaths, grass and impermeable areas such as concrete. The objective of the investigation is to record the response of various surfaces from repeated 'storms' on the same area. To ensure fair testing, always erect the simulations on gentle slopes (5–8 degrees) with approximately the same angle. Figure 2.30 shows a home-made storm simulation experiment. Steadily pour 5 litres of water into the top channel or 'cloud' and record the Initial Response Time (IRT), or the time for the water to get from the cloud to the lower channel. Using measuring cylinders, collect the runoff every 30 seconds after the IRT and discard it between each time interval. After a gap of 5–10 minutes repeat the experiment on the same surface and again record the results, indicating that this is 'Storm 2'. This can be repeated for a number of storms.
Roof gutters ▶ To measure the rainfall of runoff from the school/college/home roof. ▶ To compare rainfall and hydrographs from different rainfall events. ▶ To compare of roof/rainfall responses for different seasons.	Start by measuring rainfall (using a rain gauge) in an open location. Calculate the amount of rainfall in millimetres. Find a suitable location where the roof gutter can be accessed, and where a bucket can be inserted under the down-pipe. During a rainfall event, measure runoff amounts at frequent, regular, intervals (every 3–6 minutes depending on the intensity). From the volume of water collected, calculate the amount of runoff in litres per second. It is also possible to measure the total amount of rainfall collected from an area of roof for a particular storm. Plot the rainfall as bar charts and the runoff as a line graph on the same plot. Also measure the lag time from the time of peak rainfall to peak runoff and annotate the hydrograph components.

Infiltration

▶ To find out how infiltration rates vary on different surfaces.

▶ To consider how varying infiltration rates may affect river systems.

Many investigations seek to consider how infiltration rates vary for different surfaces. The basic factors that can be tested are: soil compaction, position on a slope and slope angle, hillslope hollows and spurs, effect of vegetation. Investigation of temporal as well as spatial differences can also be considered, i.e. repeating surveys on the same site in contrasting seasons, or immediately after a rainfall event. This type of approach might be linked to a whole range of wider catchment issues, such as flood risk for particular catchment surfaces. For different sites you could assess the likelihood of the infiltration rate being exceeded during a rainstorm (therefore surface runoff developing). Make sure that any results are plotted as hourly infiltration rates.

Runoff maps

▶ To produce a runoff map for the school grounds.

▶ To develop a (drainage) management plan for a site.

This approach involves the production of a 'runoff map' (for an area such as a section of the school grounds, or two neighbouring streets in a residential area). Obtain or produce a detailed base map of the study area. During a rainfall event make observations of where puddles form and the ground easily becomes saturated. You could score them on a sliding scale from 1–5 (1 indicating freely draining/no surface water and 5 indicating severe water-logging and standing water). The map should make reference to particular hotspots, and be colour-coded according to the degree of surface water. Also indicate the direction in which the water runs. An extension might be to suggest areas which require drainage.

Interception and stemflow

▶ To determine how much rainfall is intercepted by different types of vegetation.

▶ To consider how interception affects river systems.

Again there are a range of possibilities here. Typically experiments may involve a network of rain-gauges in open and sheltered sites, i.e. under different types of woodland (for the detection of throughfall). In the interpretation of results you should refer to a range of factors: density of canopy, time of year, leaf structure/shape, duration and intensity of rainfall, temperature and evaporation.

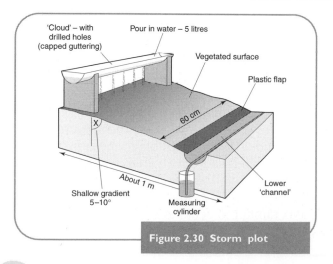

Figure 2.30 Storm plot

WEBSITES

www.environment-agency.gov.uk

www.nerc-wallingford.ac.uk/ih/www/research/iresearch.html – good source of data and information, including water quality and river flows.

www.ecn.ac.uk – again more water data, in particular data sets on environmental change.

Figure 2.31 The coastal environment

About 60 per cent of the Earth's population lives within a narrow belt of land at the ocean edge. Such a concentration of people, with the associated industrial, recreational and transport activities, creates pressure on coastal resources. Coastal systems result from the interaction of several processes including geological, weathering, erosional, depositional and biotic processes. Britain provides numerous opportunities for coastal studies with its comprehensive variety of coastal environments and scenery (Figure 2.31).

Coastal studies form ideal projects because of the dynamic nature of this environment – changes in the physical nature of the coast can happen very quickly indeed. For example, it is interesting to compare the impact of a winter storm with data collected during the summer.

The range of coastal projects can be grouped into a number of general categories (Figure 2.32). The majority of these routes to investigation are outlined in the next section. However, the ecosystem studies are found on pages 102 onwards.

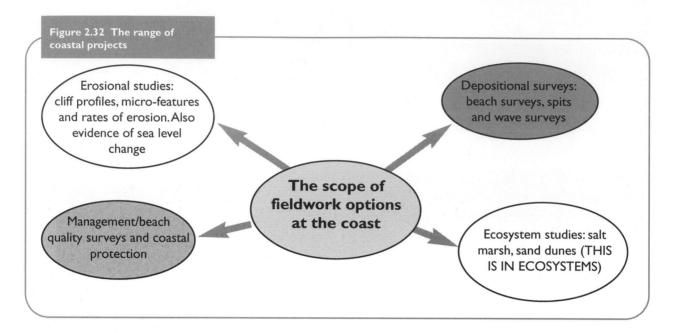

Figure 2.32 The range of coastal projects

Erosional studies: cliff profiles, micro-features and rates of erosion. Also evidence of sea level change

Depositional surveys: beach surveys, spits and wave surveys

The scope of fieldwork options at the coast

Management/beach quality surveys and coastal protection

Ecosystem studies: salt marsh, sand dunes (THIS IS IN ECOSYSTEMS)

The coastal environment

Before embarking on a piece of coastal research, there are a number of important details to consider.

▶ **Fieldwork safety.** Whatever topic you decide to undertake in the coastal environment, remember that coastal areas can be hazardous to work in. Take special note of the guidelines in the box below.

Safety in coastal areas

▶ Keep well away from cliff tops, especially in wet and windy conditions.

▶ Avoid working near the base of steep cliffs – wear a hard hat if appropriate.

▶ If you are going to measure waves, make sure you do it from a safe and secure position where you are protected from the largest (or freak) waves.

▶ Take special care on wet and slippery rocks on the beach foreshore at low tide.

▶ Be aware of the tides and never wade around a headland unless the tide is going out.

▶ **Resources.** Maps are an essential resource for many coastal projects. In particular a detailed OS map, i.e. 1:25,000, and the use of a geology map should be considered. In addition, many studies may require the use of historical maps, for instance to estimate rates of coastal recession. Make sure these are obtainable for the area in question.

▶ **Time.** Some coastal studies look at temporal variations in processes along a stretch of coast, as well as the spatial changes in particular features or systems. This means you must be prepared to return and revisit the site, perhaps to collect more data or to make more observations.

EROSIONAL STUDIES

Examples of erosional projects include the following.

Project ideas

1. How and why do cliff and shore profiles vary along the coast? For this you will need to investigate a stretch of coast, say 2–5km long, which shows several rock types and has access at different points.

2. A study of the distribution of micro-features along a stretch of coast. Again for this you will need to

---

select a length of coast which has a number of features such as stacks and arches.

3. A comparison of wave action between headlands and bays. This could be carried out on the same day, in an attempt to link wave energy to the resulting features. Alternatively you could compare wave action at different times of the year.

4. A study of recent (i.e. in the last 200 years or less) changes along the coast. This will be based both on primary data and secondary data in the form of old maps and photographs.

5. Evidence for sea level change. This will require studying a stretch of coastline with resistant geology and a predominance of features such as raised beaches and abandoned cliffs.

Cliff surveys

Working near cliffs is a potentially hazardous activity. It is essential that a hard hat is worn at all times and you would be well advised to check the time of the tides to ensure there is no risk of being stranded at the base of a cliff. Also avoid areas of the cliff which look loose and liable to slippage or collapse.

Cliffs are a recognisable indicator of a coastline undergoing erosion. Cliffs may plunge steeply into deep water or have an erosional shore platform stretching seawards from the cliff foot. The geological factors which determine the form and shape of the cliff are structure and lithology. Major structural controls include faulting and the angle of dip of the rock. Where weak material such as unconsolidated glacial or fluvial deposits are being eroded, as on the Norfolk coast in East Anglia, rates of retreat may be several metres per year and retreats of several kilometres in historical times are well documented.

Cliff surveys often involve a comparison of the **cliff profile** and its **planform**, linked to geological factors and the intensity of wave attack.

Cliff profile

Start by producing a sketch map of the cliff site, referring to a local geology map and guide if possible. Take ample photographs to support the sketch map and try to estimate the height of the cliff. This might be done in three ways:

- From a large scale 1:25,000 OS map.

- Using a ranging pole as a scale.

- Surveying with a clinometer (see page 32).

  Your sketch should be well annotated (Figure 2.33) – try to include the following details:

- **Basic geology.** Use a simple geological field guide to help in the identification of the key rock types (but don't expect to be able to recognise them all, see Figure 2.34). Look for obvious changes in the geology and mark the layers on your diagram. It may also be possible to determine the structure of the rock type in each section. Look for evidence of bedding planes (including the angle of dip), jointing, faulting and folding of the rock layers or strata. Cliff height for example, can be related to rock hardness and structure. Cliffs will tend to have a lower profile where they are made of softer rock and where the rates of accumulation are greater than the rate of removal by the sea.

- **Features.** Identify and annotate obvious features, e.g. caves, stacks, wave cut-platforms, notches, beaches, high and low tide levels and any debris which is apparent either on the cliff or at its base.

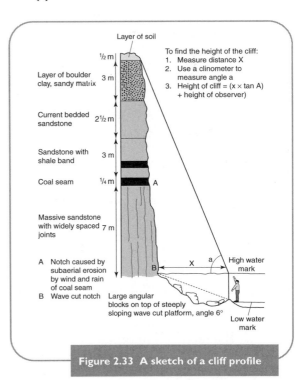

Figure 2.33 A sketch of a cliff profile

Figure 2.34 How to recognise common rocks

Increasing hardness

Chalk – When broken appears bright white. Very soft, will powder when rubbed. Often in clear layers which are easily split and weathered. Often contains black or dark coloured flints.

Boulder clay – A dry clay (brown, red or grey) found between soil and bedrock. Usually contains a mixture of other rocks of various sizes. Classified as a glacial deposit.

Sandstone – Crumbling particles and reasonably soft. Can be a range of colours from white through to yellow and brown.

Shale – These are thin layers of grey coloured rock, rather like a poorly consolidated or developed slate.

Limestone – This sedimentary rock has a number of different forms and colours, ranging from whitish to cream, yellow, brown or nearly black. It is usually fractured and jointed and may contain a number of small (sea creature) fossils. This is a fairly hard rock.

Granite – Appears almost smooth and shiny, but with speckled white, grey or pink flecks. Close viewing reveals a jumbled mass of sparkling crystals. Hard and very resistant to weathering.

▶ **Evidence of erosion.** There may be evidence of differential rates of erosion on different sections or strata of the cliff.

▶ **Mass movement and basal activity.** Look for evidence of slumping, mudslides or rock fall at the cliff base. Basal activity indicates how active cliff-foot erosion is.

▶ **Human modification/influence** Look for and comment on built structures at the base of the cliff or at the top. Is there any evidence of drainage schemes (cliff pipes), stabilisation and protection schemes?

TOP TIPS   Index of erosion and failure

You may be able to develop your own index of cliff erosion and stability for a particular stretch of coast. Use a sliding scale, where 0 indicates unstable and frequently collapsing, and 5 indicates a steep cliff which shows limited signs collapse or failure. For a short stretch of coast (1–3 km long, with a range of geologies) it should be possible to produce a colour coded and annotated map showing any areas of weakness or stability. Try to link this with a detailed geology map and comment on the strengths or otherwise of particular rock types. How important is rock type? Is slope failure controlled mainly by proximity or exposure to the sea?

Cliff plan. The planform or aerial view of a cliff coastline is strongly, influenced by geology. Wave attack, although significant, plays a less important role in the cliff's form and topography than geology. Examine the relative importance of the factors that influence the plan of the coastline at various geographical scales:

▶ **Regional scale.** At this scale OS and geological maps will show the relationship between the lie of the rocks and the broad shape of the coast.

▶ **Local scale.** At this scale, rock lithology and relative rock hardness may determine the pattern and sequence of headlands and bays.

- **Micro scale.** At this level, fieldwork might reveal that for certain lithologies and rock types, minor structural variations are responsible for small changes in the planform of the coast.

Wave surveys

Wind generated surface waves are the main source of energy along the coast. They are responsible for the erosion of the coast and for the formation of beach features. Waves can generate powerful coastal currents which cause the movement and drift of sediment along beaches.

Coastlines are classified as being either low or high energy, according to the 'normal' energy of the incoming waves. The energy of waves is related to their size and frequency. The higher and more frequent the wave, the more energetic or destructive their effect. Waves with a smaller height and a stronger swash (pushing material up the beach) are termed constructive (Figure 2.35).

A number of wave characteristics can be estimated: wave height, wave frequency, wavelength and wave energy (Figure 2.36).

Wave height is best determined by using a marker such as a breakwater or groyne; calculate the average height of 10–20 waves. You must be careful when carrying out these type of surveys, and never work alone.

Wave frequency is estimated by counting the number of waves over a 10 minute period and

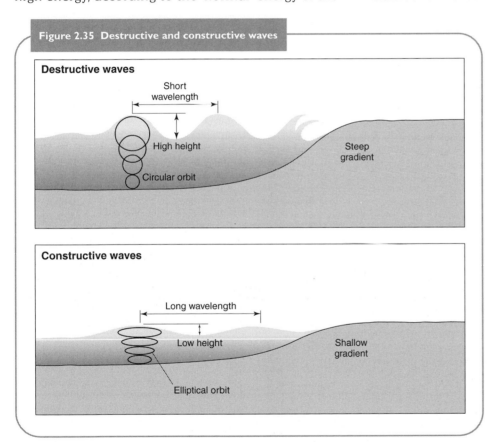

Figure 2.35 Destructive and constructive waves

Destructive waves

Short wavelength

High height

Circular orbit

Steep gradient

Constructive waves

Long wavelength

Low height

Shallow gradient

Elliptical orbit

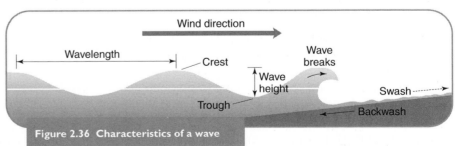

Wind direction

Wavelength

Crest

Wave breaks

Wave height

Swash

Trough

Backwash

Figure 2.36 Characteristics of a wave

dividing the total by the number of minutes to determine number of waves per minute.

Wavelength is more complex and must be calculated with a simple formula as it cannot be measured directly in the field:

For water greater than 2m deep:

$$\text{Wavelength (m)} = (1.6 \times 3600)/\text{frequency}^2$$

For water less than 2m deep:

$$\text{Wavelength (m)} = ((3.1 \times 60) \times \sqrt{\text{water depth in m}})/\text{frequency}$$

Wave energy is the amount of energy (expressed in joules per metre of wave front) that is expelled as the wave makes contact with the shore. In very rough seas the amount of energy released can be considerable and could lead to the destruction of sea defences and localised flooding of resorts. Wave energy can be estimated using the following formulae:

Wave energy (joules per metre width of wave crest) =

$$740 \times H^2 \times L$$

where H = wave height in metres, and L = wavelength in metres,

e.g. a wave of height 3m and wavelength of 40m has 266,400 joules of energy per metre width.

The above estimates are *approximate*, do not expect them to be very accurate or reliable. All these results can be put onto a recording sheet similar to Table 2.5.

An wave survey recording sheet				Table 2.5
Site/Location 24/9/01	Frequency (per minute)	Wave height (m) (H)	Wavelength (m) (L)	Wave energy (joules per metre of wave front)
Headland (water depth unknown)	22	2.6	12	50,793
	A high frequency wave, tall in height, with a big impact on the headland. High energy, probably destructive in nature.			
Bay (water depth = 1.2m)	9	0.9	22	13,186
	A lower frequency wave, providing much less energy. The wave front is also less steep. Probably more constructive in form.			

Historical erosion rates

Some parts of the British coastline demonstrate signs of very rapid coastal erosion systems. As a general rule, hard igneous rocks, for example granite, will recede at rates of less than 1mm per year. Well jointed or soluble hard rocks such as limestone and chalk may recede several centimeters in a year, while unconsolidated material will be eroded at a much faster rate. Spits, for example, indicate coastlines which are very active where there is considerable sediment movement and transfer.

It is possible to undertake studies of coastline recession rates based on past evidence. This will mostly be in the form of old maps and written records (e.g. parish documents). Primary data should also be collected in the field – there may be collapsed buildings and disused roads. These should be sketched and photographed and included within any report.

Longer-term sea level change

Evidence of the relative changes in the position of the land and sea is preserved in some areas around the British coastline. In particular, the emergent coasts of western Britain (especially in Wales and Scotland) provide strong evidence of the relative changes in the land and sea. Start by identifying a stretch of coast which is likely to show relict features such as raised beaches and 'dead' cliffs. Consult textbooks to provide named locations and examples of sea level change evidence. Some of these locations might be classified as 'RIGS' (Regionally Important Geology Sites). You should anticipate finding such features somewhere between 20 and 40m above the current sea level, although do not expect them to be very easy to spot.

In terms of fieldwork at such locations, you should concentrate on mapping features and deposits wherever possible. Use a level or clinometer to determine gradients so that a beach and cliff cross-section can be reproduced (Figure 2.37). Use detailed OS maps to determine the height of features and try to make accurate plans of any interesting evidence.

Raised beaches You will most likely find these at locations composed of resistant geology. Apart from an obvious raised beach, other clues to sea level change in the area might be:

▶ Sea shells (similar to those found on the active beach).

▶ A relic wavecut notch and wave cut platform (sloping towards the sea).

▶ Sea-caves.

▶ Rounded pebbles, possibly including some 'alien' eratics.

However, looking for, and finding this sort of evidence is difficult as many deposits and features will be covered with soil.

Abandoned/'dead' cliffs Behind the raised beach there may be a line of relic cliffs. These cliffs may show signs of previous attack by the sea (undercutting), but there will also be evidence of more recent processes acting on them, i.e. weathering and mass movements. Compare the form of the relict cliff with the current cliffs. How

and why are there morphological differences? A coastal fieldwork guide may be able to date the cliff for you, in which case you can approximate the rate of retreat.

DEPOSITIONAL STUDIES

The table below illustrates the range of depositional projects you could undertake.

Project ideas

1. How does beach form and profile vary between two different parts of the coast? Use the methods suggested on page 51 to construct two or more profiles. Try to explain any differences you find in terms of geology, exposure, wave energy etc.

2. How and why does sediment size vary across a beach? This might also include changes in sediment shape, or you might consider comparisons before and after a storm, or between two different beaches.

3. Is there a relationship between wave energy and beach form? Use the techniques described on page 60 to calculate a reliable figure for wave energy and see whether there is a testable relationship between beach gradient and/or sediment size/shape.

4. Do beaches have recognisable summer and winter profiles? The summer profile tends to be gently shelving and more sandy, whereas the winter profile is steeper and shingle-dominated. Use profiling and sediment survey techniques to test this theory for one or two local beaches.

5. An investigation of longshore drift along a stretch of coast. Using marker pebbles, keep a record of how wind speed and direction together with wave energy disturb the normal patterns of sediment transport. You will need to keep a record of all these variables, and conduct the experiment for a number of weeks.

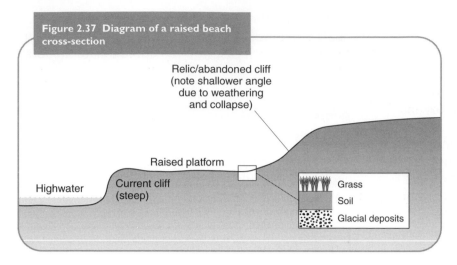

Figure 2.37 Diagram of a raised beach cross-section

Relic/abandoned cliff
(note shallower angle
due to weathering
and collapse)

Raised platform

Current cliff
(steep)

Highwater

Grass

Soil

Glacial deposits

Beach profiling

Beaches are dynamic. They seldom maintain uniform or stable slopes and are constantly being reworked and re-graded by the sea. Dramatic changes in beach form can occur especially after violent storms when destructive waves remove beach material. The first thing to do is to select a suitable area of beach upon which to undertake a series of profiles down the beach from the cliff to the sea as well as several profiles along the shore. If you are unfamiliar with the area of coast you will be working on then use detailed OS maps to help you with your site selection. Try to locate an area of beach such as a small bay where your survey can be carried out. Remember that access and safety are two

important considerations and beach profiles should be carried out at low tide.

The downshore profiles will record the different levels on the beach, such as shingle ridges, whereas the longshore profiles can indicate the effects of longshore drift (Figure 2.38). Use a clinometer or pantometer (see page 51) to measure the slope angle every 50 cm along the downshore transect. Sediment size and shape can also be determined (at less frequent intervals of perhaps 1 or 2m). One way of presenting the results obtained is shown in Figure 2.39.

It is also possible to make a more detailed and descriptive assessment of changes in the beach form from the water's edge to the cliff line or promenade (Figure 2.40). This type of approach could make a useful addition to the profiling method outline above.

Longshore drift surveys

Waves often arrive at an angle to a beach or coastline (Figure 2.41). Wave refraction causes them to be bent around, so that their final approach direction is almost parallel to the beach. This may result in the lateral shift of beach material, usually in the dominant drift direction, and the development of spits in certain locations. There are three main ways in which judgements about longshore drift can be made:

Figure 2.38 Conducting beach profiles

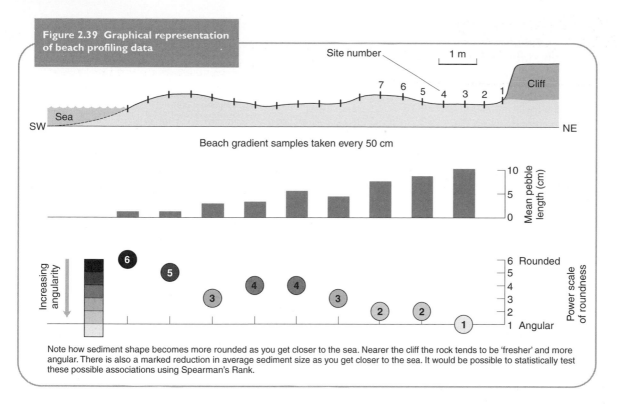

Figure 2.39 Graphical representation of beach profiling data

Site number

1 m

Beach gradient samples taken every 50 cm

Note how sediment shape becomes more rounded as you get closer to the sea. Nearer the cliff the rock tends to be 'fresher' and more angular. There is also a marked reduction in average sediment size as you get closer to the sea. It would be possible to statistically test these possible associations using Spearman's Rank.

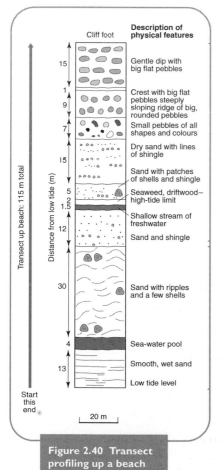

Figure 2.40 Transect profiling up a beach

▶ **Groyne measurements** This is based on differential beach heights, i.e. the difference in the height of beach material on either side of jetties or groynes which trap sediment and beach material on the updrift side. Measure (using a metre stick or similar) the height of material on either side of the groyne, making observations at the high tide, mid tide and low tide positions on the beach. If possible, measure a number of groynes at different positions along a 1 km stretch of beach. This will reveal a picture of sediment movement, especially if the recordings can be taken in different weeks or different seasons. The results can be plotted as gain/loss bar charts as in Figure 2.42.

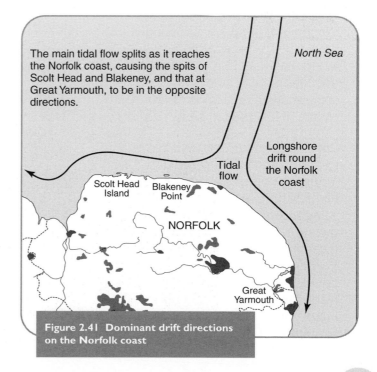

The main tidal flow splits as it reaches the Norfolk coast, causing the spits of Scolt Head and Blakeney, and that at Great Yarmouth, to be in the opposite directions.

Figure 2.41 Dominant drift directions on the Norfolk coast

▶ The direction of longshore drift will be heavily influenced by the prevailing winds – their strength, direction and frequency. This technique, therefore, might be used to investigate how important wave and wind characteristics are in determining the process of longshore drift. Using high visibility waterproof paint, spray paint a sample of stones of different sizes and shapes. You should try and use 30–50 stones if possible. Select a clear stretch of beach, avoiding any obstacles such as groynes that could interfere with longshore drift. Put the stones in the zone of swash/backwash and mark their position with a ranging pole or similar. After 20 minutes, locate the stones and determine the direction and distance travelled. You will need help to do this, and the stone recovery rate may be poor, but it is certainly worth a go. In fact, you might find that a few stones have either disappeared or travelled in the opposite direction to the majority. This experiment should be repeated a number of times to calculate an average distance travelled and to verify the normal drift direction (collect the stones and use them again). It is possible to run this experiment over a longer period of time, perhaps over days or even weeks. Be aware that you will need to use considerably more stones, somewhere in the region of 100–200 to ensure that some can be recovered.

▶ Corks (or similar) can also be used to approximate longshore drift. Use around 30–50 corks covered using a high visibility paint and use the same technique as above; marking the start point and measuring the movement after 5 or 10 minutes. Corks may be preferable to painted pebbles since they are more responsive to swash and backwash movements, although they can be susceptible to air currents.

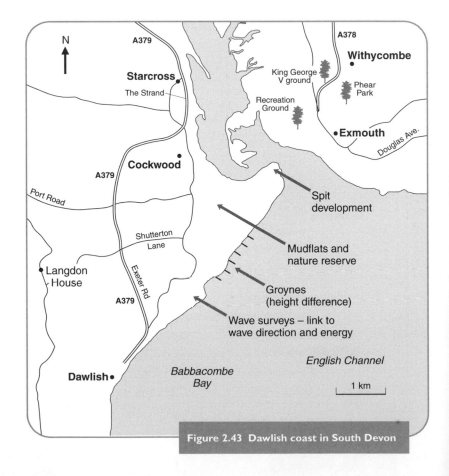

Figure 2.42 Patterns of longshore sediment movement for a 1 km stretch of beach, based on groyne measurements

Figure 2.43 Dawlish coast in South Devon

Studying processes along a stretch of coast

Depending on the stretch of coast chosen, you may be able to combine a range of techniques that have been discussed above. Figure 2.43 shows an area of coast between Dawlish and Exmouth on the South Devon Coast. For this area you can:

▶ Obtain an OS map at 1:10,000 and use this as a base map for your area of study. Annotate notable features onto your base map, including the low and high water marks.

▶ Select sites to carry out beach surveys, to produce a cross profile of the beach.

▶ Carry out wave action and pebble surveys at the points indicated on the map.

▶ Obtain an old map of the area and try to determine rates of recession.

COASTAL MANAGEMENT SURVEYS

Since coasts are often extensive areas of flat, fertile, picturesque land, such environments are very attractive for a wide range of human activities. Approximately 25 per cent of the English coastline has been heavily developed for housing, industry, agriculture and leisure and is always under attack from both the local weather and the action of the sea, resulting in coastal erosion.

For centuries this human intervention along the coastline has modified the effects of natural coastal processes. Measures have been taken in many locations to protect urban areas and low-lying agricultural land from flooding and erosion by the sea. Coastal defence structures such as sea walls, groynes and breakwaters are widespread and obvious. The replenishment of beaches and the stabilisation of dunes are examples of more recent, less obvious defence techniques.

Management surveys and the range of themes that this subject encompasses often involve real and tangible issues that affect both people and places. Coastal management issues are also now gaining a higher political profile, thus increasing their appeal to the enquiring geographer. There are a number of ideas that deserve consideration, and these can be summarised under three broad headings: coastal protection surveys; tourism, amenity and activity surveys; and beach and/or water quality surveys.

Coastal protection surveys

The more coastal areas are developed, the greater the need to protect and defend them from destructive wave energy and coastal flooding. This provides opportunities to investigate the effectiveness of such schemes.

Project ideas

1. **Defend or retreat?** This style of project should evaluate the effectiveness of coastal management in the area. It might also attempt to comment on the likely success of the scheme in the future. A range of techniques could be applied: beach profiling and sediment analysis, evaluation of protection scheme questionnaires, photographs and sketching. Cost-benefit analysis might also improve the quality of data analysis.

2. **Do we need a coastal protection scheme?** This investigation might be based on the erosion or flooding problems in a particular area and should focus on the views of local interest groups. A number of possible options should be reviewed and evaluated. The data collection should be centred around a comprehensive set of questionnaire responses from local residents, businesses and tourists. In addition to these, primary data must also be collected to support the degree of coastal erosion (or retreat) and/or flooding. Use photographs, sketching and other suitable methods to support your findings. Look for erosion in the form of slumping, cliff cracking, undercutting etc.

Firstly select a coastal strip which contains a range of protection schemes, where the intensity of land usage varies from very high (e.g. resort towns) to low (e.g. caravan parks and golf-courses). Complete a land use map which is colour-coded by function (an example land use key can be found on page 133). The length of coast selected should be somewhere between 2 and 5km. Research archive material such as historical photos, newspapers and maps to get an indication of changes in the coast over time.

Carry out your own survey of the existing coastal protection schemes, using the following criteria: cost, condition, effectiveness and visual quality. If you use a sliding scale, i.e. from 1–5, to quantify the criteria, support your data collection with photographic evidence. Table 2.6 identifies the range of coastal protection surveys commonly used. Evidence of coastal erosion should also be measured, such as slumping of cliffs, obvious notches, water coming from the cliff, undercutting and cliff cracking. Abandoned roads (from OS maps) and land/property prices can also be used as an indicator of the severity of coastal erosion.

Coastal protection schemes — **Table 2.6**

Coastal scheme	Typical location and function	Approximate costings	Typical lifespan (years)	Possible outcomes
Concrete sea walls	Popular sea-side resorts. Double up as promenades.	£5000 per metre	50–75	Deflect waves downwards and increase beach erosion, eventually undermining the sea wall.
Revetments	Lower cost defences along less developed coastlines.	£2000 per metre	Less than 50 years	Effective at breaking and absorbing wave energy, but do not give complete protection.
Groynes	On tourist beaches, spaced every 100–200.	£10,000 each	25–40	Trap and maintain sand on tourist beaches, but may lead to sediment starvation downdrift.
Gabions	Boulder-filled wire cages often behind sea walls or revetments.	£500 per metre	10–30	Rapidly break up if exposed to high wave energies. Over time they can be stabilised by vegetation.
Earth embankments	Flood defences along low-lying, low intensity coastlines.	£1000 per metre	Variable	Protect farmland, but prevent sea from encroaching inland, finding its line of equilibrium.
Rip-rap	Huge piled boulders, patching up earlier failed schemes.	£1000 each rock	Short term	Huge mass prevents movement in storm waves, gaps absorb wave energy.
Beach feeding	Gravel/sand dumped by local authorities on tourist beaches.	Cheap	Short term	Easily removed by offshore currents and longshore drift.

A questionnaire survey of local residents and visitors to the area can be undertaken to determine their awareness of coastal protection (Figure 2.44). In particular, responses may reveal different user groups' perception of:

▶ The coastal threat.

▶ The response by the authorities.

▶ The impact of the coastal defences.

Proposed coastal protection schemes can also be evaluated using a simple cost-benefit analysis. The aim is to view the total cost of the scheme over its life-span, compared to the costs that would have been incurred had the scheme not been implemented (benefits). Table 2.7 is a simplified cost-benefit table. In reality it is very difficult to get real data on costs and benefits, so you will have to expect to use best guess estimates based on your own (justified) figures.

Coastal cost-benefit — **Table 2.7**

Costs	Benefits
Construction of scheme	Protection of infrastructure
Ongoing maintenance	Protection of farmland/ buildings
Purchase of land	Protection of tourist amenities, e.g. beaches
Compensation	Reduction of flood risk
Loss of visual quality	'Peace of mind' for residents
Knock-on effects	Reduced cost of emergency services

Figure 2.44 A student questionnaire designed for residents of an area with a coastal defence scheme

'A' LEVEL GEOGRAPHICAL SURVEY:
SIDMOUTH COASTAL DEFENCE SCHEME.

Please tick the appropriate boxes:

1. Age:
0–20 ☐   41–60 ☐
21–40 ☐   61+ ☐

2. How long have you lived in Sidmouth?
0–5 years ☐   11–20 years ☐
6–10 years ☐   21+ years ☐

3. Have you heard of this proposed scheme?
YES ☐   NO ☐

If yes, do you consider the scheme to be better than previous attempts to stop the erosion on Sidmouth beach?
YES ☐   NO ☐   DON'T KNOW ☐

4. Do you think that the scheme is worthwhile in terms of preserving the beach, or could the money be put to a better use i.e. cleaner beaches?
VERY WORTHWHILE ☐   WORTHWHILE ☐   NOT WORTHWHILE AT ALL ☐

5. Which of the following impacts would you consider the constructive phase of the scheme to affect the most? (Please rank 1–5, with 1=most affected and 5=least affected)
VISUAL ☐
NOISE ☐
ENVIRONMENTAL ☐
TOURISM ☐
LOCAL RESIDENTS ☐
TRADE ☐

6. How will it affect you
NOT AT ALL ☐   A LITTLE ☐   VERY MUCH ☐   NOT SURE ☐

7. Do you think that Sidmouth will have the same appeal in the future as a result of the scheme, i.e. to tourists, potential newcomers to the area, and indeed the established residents?
YES ☐   NO ☐

Additional comments: _____
_____
_____

THANK YOU VERY MUCH FOR YOUR CO-OPERATION

TOP TIPS

Shoreline Management Plans (SMPs)

A Shoreline Management Plan (SMP) provides a large scale assessment of the risks associated with coastal processes and presents a policy framework to reduce these risks to people and to develop the natural environment in a sustainable manner. SMPs are 'working documents' for a specified length of coast, and are implemented by the relevant coastal authority. There are four main strategies usually available for a stretch of coast:

1. **Do nothing** carry out no defence works – this could lead to continued erosion or flooding.

2. **Hold the line** hold the line of defence in its current position.

3. **Advance the line** move defences seawards – this could result in loss or degradation of intertidal habitats.

4. **Retreat the line** move the existing defence line landward (i.e. managed retreat). This could lead to loss or destruction of freshwater habitats behind current defences, especially in the case of coastal sand dune systems.

You could use the guidance set out in an SMP to assist in the development of your own management plan for an area of coast. This might form a project in its entirety, or part of a wider study of the coastal area. Figure 2.45 shows an example of a free downloadable map from the Solent SMP internet site. The references at the end of this section provide some additional internet links, including a downloadable summary of the SMP operation from the DEFRA site. Full SMPs are weighty documents, so be careful not to get 'snowed under' with the detail.

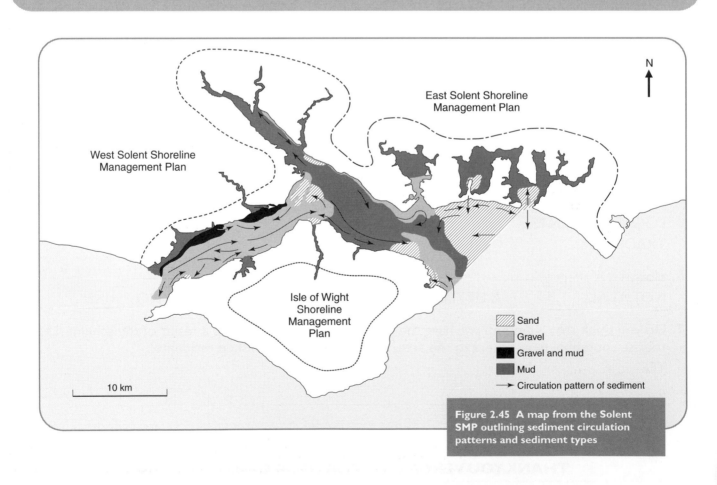

Figure 2.45 A map from the Solent SMP outlining sediment circulation patterns and sediment types

Tourist/amenity surveys at the coast (see also Sport, Leisure and Tourism, p. 218)

There are a whole host of ideas and topics that can investigated within this heading.

Project ideas

1. How and why activity 'hot-spots' change during the course of a day for a named resort. Develop base maps for the resort and conduct a pilot survey to establish likely 'hot-spot' areas. Plot people's patterns of movements and try to understand any spatial and temporal patterns. This could be extended to identify areas which have parking and traffic problems. What might be the solutions?

2. A comparison of the seasonal patterns of usage along a stretch of coast. This is a longer-term study, so access to the coast is essential. Again, use two or more resorts and try to establish how patterns of tourism vary. Questionnaires may reveal patterns of activity, type of accommodation and length of stay. Are there significant differences between low and high season?

3. Is tourism a necessary evil? A survey of the impact of tourism on a coastal resort. Different people will have different views about tourists in any one area. For many coastal areas it will be vital to the economy. What proportion of local shops are tourist orientated? Are there any seasonal shops for tourists. Use questionnaires to find out the views of local residents and businesses.

4. A study of the sphere of influence for a coastal resort. This might be based on the assumption that the larger a coastal resort, the more extensive its sphere of influence. Select two or more suitable resorts and use questionnaires to determine the sphere of influence. How and why does it vary? Use census data to establish populations and OS maps to work out the area of each settlement.

Usually these types of survey are used to enable comparisons to be made between different coastal locations, or the same resort at low and high season, or at different times of the day. It may be worthwhile to compare different parts of the same resort to identify traffic and tourism 'hot-spots' for example, or to make comparisons between different resorts. This can be accomplished with the use of OS base maps, land use maps, traffic flow surveys and pedestrian density flows. For this reason, these surveys are well suited to a group approach so that large quantities of data can be collected.

Also consider the construction of an 'impact matrix' (Table 2.8) to subjectively assess how

A coastal resort impact matrix — Table 2.8

✗ = negative mix
? = uncertain impact/mix
✓ = peaceful/positive mix or no impact

	Swimming	Sunbathing	Dog walking/walking	Beachcombing	Beach games	Sandcastles	Fishing	Running	Watersports	BEACH QUALITY	ENVIRONMENT	COMMUNITY
Swimming	–	✓	✓	✓	✓	✓	✗	✓	✗	✓	✓	✓
Sunbathing	✓	–	?	✓	?	✓	✓	✓	✓	✓	✓	✓
Dog walking/walking	✓	?	–	?	✗	✗	✓	✓	✓	✗	?	✓
Beachcombing	✓	✓	?	–	✓	✓	✓	✓	✓	?	✓	✓
Beach games	✓	?	✗	✓	–	✗	?	✓	✓	?	✓	✓
Sandcastles	✓	✓	✗	✓	✗	–	✓	?	✓	✓	✓	✓
Fishing	✗	✓	✓	✓	?	✓	–	✓	✗	?	?	✓
Running	✓	✓	✓	✓	✓	?	✓	–	✓	✓	✓	✓
Watersports (includes motor sports)	✗	✓	✓	✓	✓	✓	✗	✓	–	?	✗	?

activities interact with each other, the local community and environment.

There are a number of other amenity surveys that you might consider at the coast, including the effects of recreational pressure. This is often very apparent in sensitive environments such as sand dunes and salt marshes, and the impact of trampling is important on coastal footpaths.

Beach and water quality surveys

Often a focus for these types of surveys is pollution: recording (and using photographic evidence) to note such things as chemical froth, oil pollution and sewage contamination. This could be done systematically along a series of transects from the sea to the promenade or cliff base. Daily sea temperatures can be monitored with a thermometer. Secondary data on beach and water quality can be obtained from your local authority or regional Environment Agency office. Consider a comparison of winter and summer bathing water quality and sea temperatures, or a comparison of two local resorts. Remember that much of the climate and water quality data should also be displayed in local newspapers. Your local Environmental Health Officer might also be a good person to contact with respect to beach and water quality. They might be able to give you information relating to effluent discharge points and the degree of water treatment prior to discharge into the sea.

Project ideas

1. A study of the beach quality in two adjacent coastal resorts. Use your own criteria to establish a quality index. Measure the beach width and produce a map which shows the type (and possibly size) of beach material. Does quality vary on a day to day basis, or is it more seasonally pronounced? What additional controls are there on beach quality? Are people's beach activities influenced by beach quality?

2. How and why does bathing water quality vary during the course of the summer season? Use simple observations of litter, rubbish, oil pollution etc. to qualitatively score beach quality. What controls are there on quality? Is quality worse or better after a storm? Mark your findings on a base map and support them with photographs.

3. A survey of beach facilities along a stretch of coast. Mark the beach facilities on to a well annotated and detailed base map. Try to establish whether there is a link between the number of visitors and facilities, and whether this controls the catchment of the beach. A questionnaire may reveal whether people are more interested in the beach quality, i.e. amount of sand and beach width, or other facilities such as toilets, ice-cream shops and amusement arcades.

These types of investigations might be broadened to include a wider review of beach quality and amenities, in particular the following criteria are suitable for evaluation:

▶ Amount of beach litter and provision for litter and recycling.

▶ Ease of access to the beach – car parking, bike racks, footpaths etc.

▶ Quality of access for people with mobility problems.

▶ Quality of beach material, i.e. proportions of sand and rocks.

▶ Provision of lifeguard facilities and a first-aid point.

▶ Bathing safety, i.e. slope of beach into the water, rip tides etc.

▶ Facilities – toilets, shops etc.

▶ Seaside activities, i.e. swimming, surfing, canoeing, snorkelling/diving etc.

Make a detailed base map of the coastal resort and mark on basic facilities as indicated from the criteria above. It could also be colour coded according to land use. Use an activity survey to identify the most popular areas of the beach and see if this is where most attractions are found. Is there a distance decay of tourists away from the central areas of the beach? Are the most crowded areas of the beach those which experience the best or worst beach quality? Bathing safety can be determined by establishing the gradient of the beach into the sea, and by consultation of tide tables to determine the heights (in metres) of the tides. Generally more exposed and therefore potentially more dangerous beaches have greater tidal variations.

TOP TIPS
Blue Flag beaches

The EU officially classifies dirty beaches as having an excess of 2000 faecal coliform bacteria per 100ml of bathing water. This information is displayed on all tourist beaches. Some beaches may receive a 'Blue Flag' Award if they meet the criteria shown below.

Blue Flag and bathing water quality

This beach is a Blue Flag beach.
It means that a number of requirements regarding bathing water quality, cleaning of the beach, toilets, safety and information and environmental activities are fulfilled.

It also means that the bathing water is continuously monitored for the three different types of bacteria shown in the tables. The bathing water is monitored at least every fortnight throughout the season.

In the tables you can see when the water has been analysed and how many bacteria were found.

A small number of bacteria tells you that the bathing is very clean - a high number of bacteria tells you that the water may be polluted and could contain bacteria from sewage. The limit for how high the numbers are allowed to be are shown to the right.

WEBSITES

www.rigs.org.uk – regionally important geology sites. May help you to track down raised beaches and abandoned cliffs

www.old-maps.co.uk – old maps from 1890, 1900

www.solentforum.hants.org.uk – downloadable SMP for the Solent area. Note the full document, including maps etc. is large at 7MB

www.defra.gov.uk/corps/consexer/shore/shore1.pdf – a downloadable pdf file outlining the processes involved in the development and implementation of SMPs

www.coastalguide.org – an extensive site about the European coastline, including information on habitat types and conditions, also downloadable maps

www.blueflag.org – the European Blue Flag website

www.blueflag.org/criteria/beachcriteria.doc – a detailed document providing information about the Blue Flag Award

www.goodbeachguide.co.uk – resort-based guide, providing detail on facilities and location maps (OS) for a range of UK beaches

ICE DETECTIVE

Figure 2.46 Did it once look like this?

You will have to use various sources of evidence to discover the scale and nature of glaciation in the study area

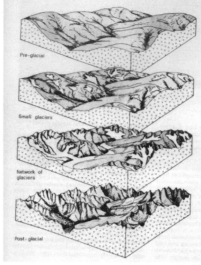

Pre-glacial

Small glaciers

Network of glaciers

Post-glacial

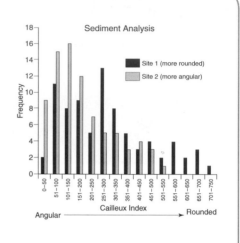

Sediment Analysis

■ Site 1 (more rounded)
□ Site 2 (more angular)

Frequency

Angular ——————→ Rounded

Cailleux Index

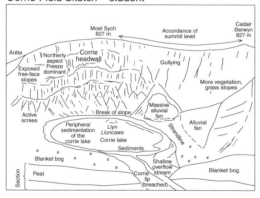

Corrie Field Sketch – student

Moel Sych 827 m
Cadair Berwyn 827 m
Accordance of summit level
Arête
Northerly aspect Freeze dominant
Corrie headwall
Gullying
Exposed free-face slopes
More vegetation, grass slopes
Active screes
Break of slope
Massive alluvial fan
Peripheral sedimentation of the corrie lake
Llyn Lluncaws Corrie lake
Alluvial fan
Strandline
Sediments
Blanket bog
Shallow overflow stream
Blanket bog
Section
Peat
Corrie lip (breached)

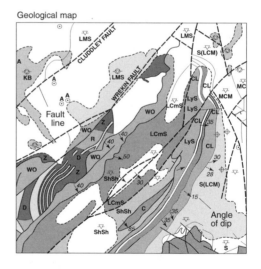

Geological map

CLUDDLEY FAULT
WREKIN FAULT
LMS
S(LCM)
LMS
A
KB
A
A
LMS
CL
CL
MC
LCmS
LyS
MCM
Fault line
WO
WO
LCmS
LyS
CL
R
Z
40
40
50
LyS
CL
Z
WO
30
WO
D
28
ShSh
30
S(LCM)
D
40
C
15
LCmS
35
35
ShSh
ShSh
35
Angle of dip
35
S

British landforms display ample evidence of processes which were more active in past climates (Figure 2.46). In fact the bulk of the present landscape consists of relic forms, largely inherited from processes operating in cooler conditions which dominated the last 2 million years of Earth history.

Glaciated landscape regions show impressive morphology which results from past glacial and periglacial processes and can provide ideal environments for fieldwork. When you are choosing a glacial or periglacial project you will need to be a good 'Ice Detective' as in Britain all of the ice disappeared approximately 10,000 years ago. You will need to do a lot of research about present day glaciers in other regions so that you understand how they worked in the past. Your investigation will probably be based on a relic landscape, where some of the major landscape features were either formed or modified by ice, but more recently they have been further changed by the action of rivers and/or weathering.

ICE PROJECTS

The range of projects under this heading includes the following.

▶ The patterns of distribution of major erosional or depositional features (these will be relic features).

▶ Studies on how and why the ice moved.

▶ Analysis of sediments – their size, shape and orientation can reveal something about the nature of the environment in which they were deposited.

Alternatively, there are also a number of possible projects which relate to people's use of the glaciated landscape – tourism, amenity and activity studies.

The next sections are organised under the following main themes:

Survey technique	Use and relevance
1. Distribution and orientation of features	This is a relatively straightforward option and includes the use of both primary field evidence and detailed map evidence. Typically the distribution of features such as corries and drumlins are studied, together with their orientation and size.
2. Morphological mapping	An upland environment is probably most suitable for this type of project, but it may be possible in a lowland area which shows significant relic features. You should map the location of large blocks and boulders, orientation of features, e.g. drumlins and eskers, evidence of striations or the distribution of moraine across a valley.
3. Provenance studies	You can also look at where the ice came from. The size, position and orientation of features can reveal something about their past, together with the type of ice that was responsible for their development/evolution.
4. Sediment studies	There are a number of sediment characteristics such as sediment size, shape, orientation and degree of sorting that can be analysed in order to understand the movement of ice in an area and to link current and past processes to the present landforms.
5. Weathering studies	Studies of weathering may reveal something about different aspects and altitudes within contrasting mountain environments. Although unlikely to provide the basis for a study in its own right, this approach might be used to complement other approaches.
6. Alpine projects	Some schools now visit areas such as the Alps and Iceland so there are a range of options, including the human effect of ski tourism on the physical landscape and its environmental impacts.
7. Activity and amenity surveys	These projects can be related to sphere of influence, activity pressure points etc. in areas such as the Lake District and North Wales.

DISTRIBUTION AND ORIENTATION OF FEATURES

'Distribution' type studies offer relatively straightforward projects, utilising a combination of both primary field evidence and secondary resources such as maps. The fieldwork should include a visit to at least two sites, or a location where two example features can be investigated, measured and photographed. Then, using detailed OS maps at 1:25,000 scale, note the distribution of other similar features. You should aim to locate 10–15 in the same region if possible. Record the six-figure grid reference of each located feature for your distribution analysis and construct a simple distribution map if relevant.

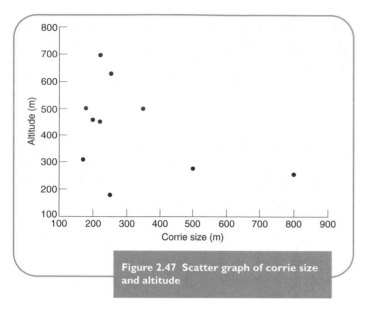

Figure 2.47 Scatter graph of corrie size and altitude

Upland environments

There are a number of potential topics for study involving corries in mountain environments.

▶ **Is there a relationship between corrie size and altitude?** Corrie sizes can be measured in the field (by pacing), or by establishing the area of a tarn if the corrie has one. To determine corrie height from a map use a large scale 1:25,000 version. If you have access to a global positioning system (GPS) device, this can provide accurate altitude measurements as long as it is correctly calibrated. The results can then be plotted as a scatter graph (Figure 2.47). Providing there are sufficient data pairs (at least 8), consider using Spearman's Rank correlation test statistic.

▶ **Is there a relationship between corrie size and the steepness of the backwall?** Again, use either field or map evidence (or a combination of both) to determine corrie area, and use a clinometer/map data to establish the backwall gradient.

You should take a number of measurements to find a reliable mean.

▶ **What is the preferred direction/aspect of the corries in area X compared to area Y?** Use a compass to determine aspect in the field, alternatively aspect can be estimated from a large scale OS map (1:25,000 or larger). Use two areas

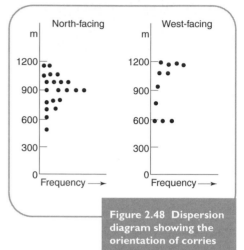

Figure 2.48 Dispersion diagram showing the orientation of corries

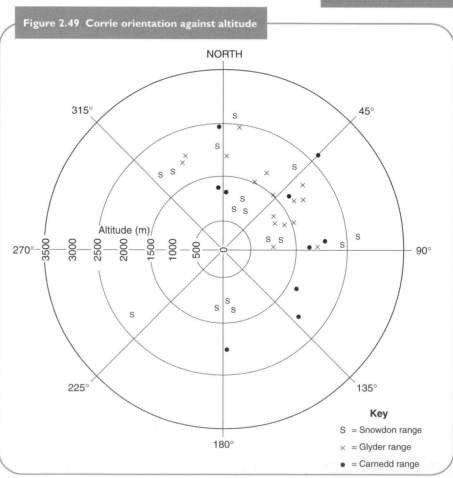

Figure 2.49 Corrie orientation against altitude

Key

S = Snowdon range
× = Glyder range
● = Carnedd range

in a similar region or compare contrasting locations (Figure 2.48).

▶ **Is there any pattern in corrie orientation and altitude?** Determine the aspect, orientation and altitude of each corrie and the results can be plotted as a radial graph as in Figure 2.49.

▶ **Is there a relationship between size/area of a tarn and the exit discharge of lake streams?** Calculate the area of the tarn under investigation (either field or map based). Then calculate the discharge of each stream. The discharge is the cross sectional area of the channel multiplied by the velocity of the water (see page 33). Plot your results as a scatter graph and consider using Spearman's Rank to establish a correlation coefficient and therefore test any hypotheses or null hypotheses.

drumlin has three components: the gradient of the stoss (steep) side, the angle of the tapering end and the angle of the sides (Figure 2.50). Use basic trigonometry and/or graph paper to plot the shape.

Elongation ratio

You can try out the 'elongation ratio' on individual drumlins:

7:1, i.e. the width is 7 times the length. Does it work?

Lowland areas

Much of the lowland regions within Britain show evidence of glaciation which has affected them in the last 100,000 years. Usually the landforms which you will find in lowland areas are those of glacial deposition. The features are likely to be small scale and relatively difficult to identify from basic map evidence alone.

▶ **A study of the form and distribution of drumlins** A drumlin field or swarm is often said to have 'basket of eggs' relief because of the streamlined topography of each drumlin. You will need to make a detailed survey of a group of drumlins, including their size, shape, distribution and orientation.

The length of the drumlin can be determined by using a tape measure or accurate pacing. The height can be determined with a clinometer. The shape of the

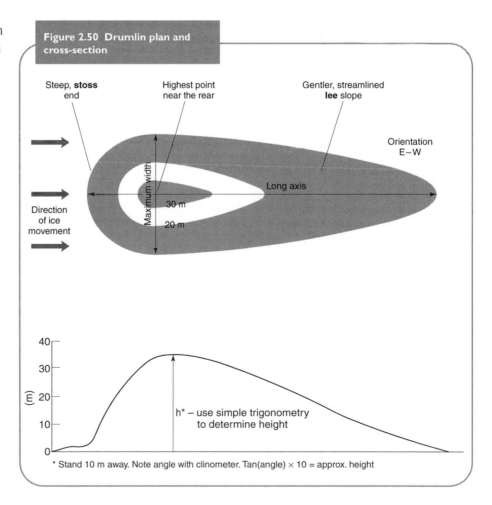

Figure 2.50 Drumlin plan and cross-section

Using a large scale map plot the distribution of the drumlins or measure the distance between the highest point of each one and its nearest neighbour. This simple statistical test can then be used to reveal the pattern and distribution of the drumlin field. You may well find that there are differences in the spacing and distribution of drumlins between contrasting areas. Try to establish a possible explanation for this.

Use a compass to find the orientation of each drumlin (along its long axis). The orientation can be used to indicate the general direction of ice flow. Again, plot the orientation of drumlins for contrasting areas and try to account for any differences that you might find.

If you can find an exposure through a drumlin, sketch the internal structure and sediments. Try and identify what the core is composed of – does it have an identifiable core and till shell covering it?

Other 'distribution' type studies

If you have access to an area containing roche moutonnées or crag and tails, attempt to measure the former ice flow direction using a compass, remembering to note the stoss and lee sides of the features. It may also be possible to calculate the density of fractures on the surface of the feature by setting out a grid and counting the number of fractures per square metre (or the area of the grid/quadrat). Is the roche moutonnée form strongly influenced by rock structure?

Using 1:25,000 maps for reference, choose an area which shows evidence of past glacial erosion, e.g. frequent glacial lakes. Use the density of these lakes (per square km) as an indicator of glacial erosion intensity in the area. Compare different areas and see whether these are significant differences. Use a local geology map to investigate if the intensity of erosion relates to the regional structure pattern, i.e. density of faults and intrusions.

MORPHOLOGICAL MAPPING

This technique refers to the detailed examination of a landscape and the recognition of how ice has modified the features within an environment. The

basis for this type of approach should be to determine how and why the presence of ice has changed the landscape which can be seen today. This kind of approach can be used in both lowland and upland areas, although it is preferable to choose a small scale area, perhaps a few square kilometres, where field evidence can be easily collected.

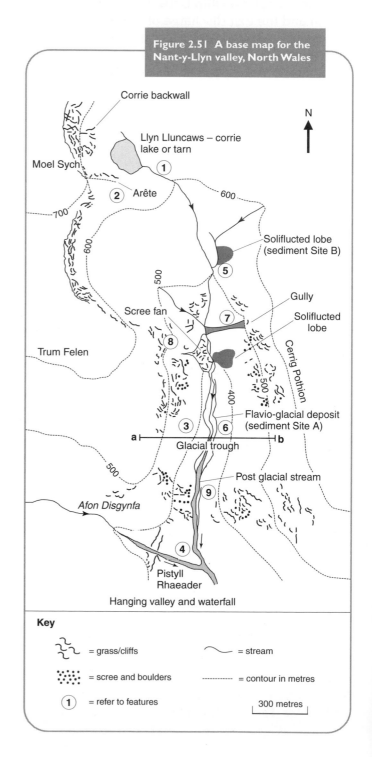

Figure 2.51 A base map for the Nant-y-Llyn valley, North Wales

Corrie backwall

Llyn Lluncaws – corrie lake or tarn ①

Moel Sych

② Arête

— 700 —

— 600 —

N

— 600 —

— 500 —

— 700 —

Soliflucted lobe (sediment Site B)

⑤

Gully

Soliflucted lobe

Scree fan

⑦

⑧

Trum Felen

Cerrig Pothion

— 500 —

— 400 —

Flavio-glacial deposit (sediment Site A)

③ ⑥

a ├─────────────┤ b

Glacial trough

— 500 —

Post glacial stream

⑨

Afon Disgynfa

④

Pistyll Rhaeader

Hanging valley and waterfall

Key

〰 = grass/cliffs

∴ = scree and boulders

① = refer to features

⌇ = stream

----- = contour in metres

├──────┤ 300 metres

Start with the construction a base map of the area, derived from a large scale, 1:25,000 OS map (from which most glacial features can be recognised). Any field guides or past projects carried out in the area should also be researched.

Figure 2.51 is an example of a base map for the Nant-y-Llyn valley in North Wales. Note how the map is labelled and has numbers which refer to particular features. In addition, a booking sheet should be designed so that appropriate data can be easily gathered in the field (Figure 2.52). The key to a successful project of this type is the explanation of what you have found. Why are particular features found in certain areas? How has the area been modified in the last 10,000 years since the ice disappeared?

Glacial Features Booking Sheet

Valley name:
Date of survey:

Orientation of valley: . . . .

Length of valley: . . . .   Valley floor gradient: . . . .

Across floor width: . . . .   Across floor gradient: . . . .

Height of valley side in section A–B: . . . .   Gradient: . . . .

Height of valley side in section C–D: . . . .   Gradient: . . . .

Features:
(Letters refer to positions on the sketch map)
Hanging valleys:

Corries:

Figure 2.52 Glacial features booking sheet

TOP TIPS

You might consider a more detailed survey of a glacial area, using surveying techniques to determine and reconstruct long and cross valley profiles. Also provide detailed field sketches as in Figure 2.53; this will reveal to the reader of your project your understanding of the past and present processes. Always complement any field sketches with high quality annotated photographs.

Linked to this type of survey are studies of **provenance**, i.e. the origin of a particular glacial feature, which in turn helps you to understand more about where the ice came from. In particular you might choose to map and study one of the following.

▶ The location and distribution of large boulders. How did they get there? Was it ice that placed them in their current position or more recent mass movement? This could include the study of erratics (i.e. rock particles or blocks transported by ice which are of a different geology to the country rocks of the area in which they have been deposited). You might even attempt to calculate the mass of these blocks, and therefore the amount of energy required to move them (clue: rock has a density about 2.5 times that of water, so a cubic metre of rock = 2.5 metric tonnes).

▶ The location, distribution and abundance of moraine within an area. Try to categorise the different types of moraine that might be present, i.e. medial, lateral, terminal etc.

▶ If you have access to glacially-eroded bedrock surfaces, measure the directions of the striations (scratches produced by rock embedded within the ice as it moves over a surface). Note down any cross cutting evidence. Is it possible to identify any changes in glacier flow? Be sure to support your evidence with well annotated photographs.

▶ The orientation of features such as drumlins and eskers may also provide clues as to the direction of ice movement.

Figure 2.53 Field sketch of the Nant-y-Llyn valley

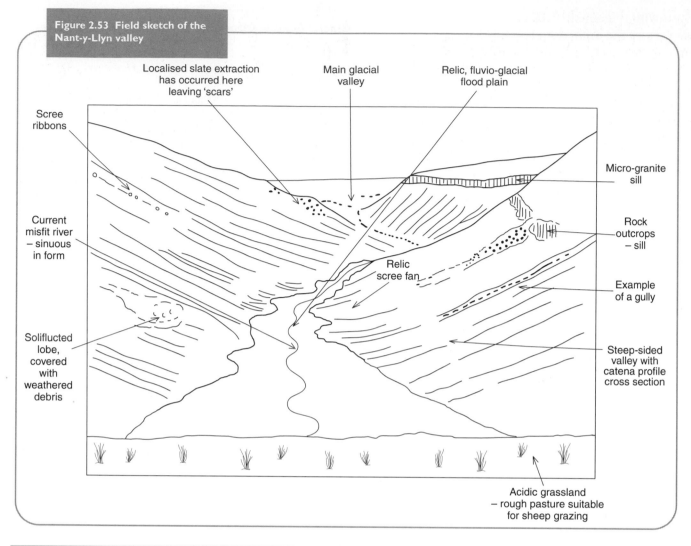

Localised slate extraction has occurred here leaving 'scars'

Main glacial valley

Relic, fluvio-glacial flood plain

Scree ribbons

Micro-granite sill

Current misfit river – sinuous in form

Rock outcrops – sill

Relic scree fan

Example of a gully

Soliflucted lobe, covered with weathered debris

Steep-sided valley with catena profile cross section

Acidic grassland – rough pasture suitable for sheep grazing

SEDIMENT ANALYSIS

Fieldwork in relic glacial landscapes is often centred on the examination of sedimentary deposits. These can contain a wealth of information: the origin of a deposit, its mode of formation and possibly the environment under which it was deposited. However, even relatively simple field exposures may initially appear to be of such bewildering complexity that they defy rational explanation. In addition to this, sediment analysis is one of the least glamorous of all fieldwork techniques, requiring patience and the skill to take repeated measurements without complaining!

Sediment analysis provides a small window into the past processes and conditions dominant in the area. When examining a sediment or series of sediments, the key information required relates to the:

▶ size

▶ shape

▶ extent of sorting of the particles or material.

This may reveal how the material was transported and deposited, i.e. by ice or meltwater. Stones and sediments in glacial environments will be irregular in shape and size, thus presenting problems in terms of their measurement.

Your basic starting point should be a geological map and if possible, a glacial guide for your chosen area. To find a relevant sediment you may have to dig into an unidentified feature but some sediments may be relatively easy to come across, being located in exposures such as river cliffs (Figure 2.54), which makes subsequent extraction and analysis more straightforward.

Figure 2.54 River cliff showing glacial deposits

Depending on the type of study, ensure that at each sediment site a short description is taken, along with an accurate six figure grid reference and photograph if possible. It is also useful to determine from a large scale OS map the approximate height and aspect of sediment sites. If you can get access to a handheld GPS, this will assist with location and height descriptions considerably.

Sediment size

Sediment size is one of the most useful parameters, but potentially complex to survey as there are three possible axes that can be determined (Figure 2.55) .

TOP TIPS — How many sediments?

There are no hard and fast rules as to how many sediments to measure; however the number of sediments measured; should be determined by considering:

▶ The number of sites that will be surveyed.

▶ How much time you are devoting to this aspect of the fieldwork.

The more stones that can be measured the better (or more reliable) your results will be. Aim for a *minimum* of 50 at each location.

If the stone is not too irregular, the length of the middle axis is most strongly correlated with overall size, making it the most representative single measurement. Sediments of over about 5mm radius can be measured with a ruler or calipers.

There are also devices, variously named 'pebbleometers' which can be used to make an assessment of sediment size (Figure 2.56). Unless your school/college has one, this can be home made from either wood or aluminium. The pebbleometer consists

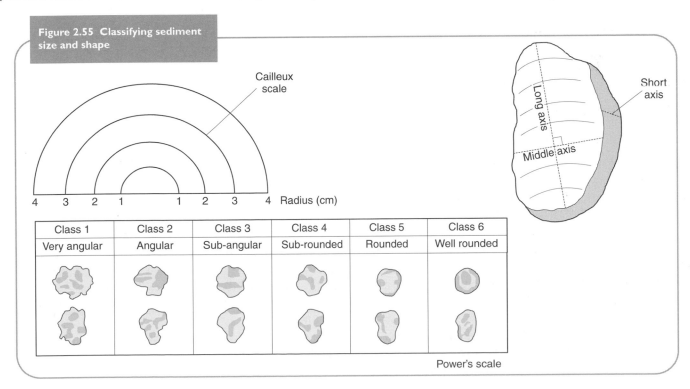

Figure 2.55 Classifying sediment size and shape

Cailleux scale

4 3 2 1 1 2 3 4 Radius (cm)

Long axis

Middle axis

Short axis

Class 1	Class 2	Class 3	Class 4	Class 5	Class 6
Very angular	Angular	Sub-angular	Sub-rounded	Rounded	Well rounded

Power's scale

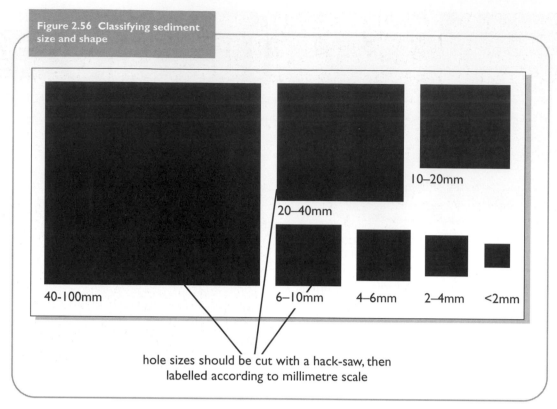

Figure 2.56 Classifying sediment size and shape

10–20mm

20–40mm

40-100mm

6–10mm

4–6mm

2–4mm

<2mm

hole sizes should be cut with a hack-saw, then labelled according to millimetre scale

of a series of different sized squares, each one representing a 'phi' value, or category of class size (see Table 2.9 and Top Tips box). They have a number of advantages over the traditional ruler method:

- They are easy to use so that large samples can be quickly generated (especially when working in groups).

- They measure the middle axis which is the most useful parameter.

- The type and nature of the deposit can be inferred.

Clearly it is not easy to measure stones with diameters less than 2–3mm by using either a pebbleometer or ruler. The sizes of these particles have to be determined by sieving (see page 109).

Sediment shape

Sediment shape is used to indicate something about the nature of the environment in which the sediments were deposited. Sharp or angular material is associated with direct ice contact, whereas more rounded sediments indicate the action of abrasion and attrition by (melt) water.

There are three main field survey techniques for determining sediment shape:

- **Powers Scale** This offers a quick (but subjective) assessment of sediment shapes based on a visual comparison. Simply compare each stone or pebble in your sample and note the number in each category (see Figure 2.55).

- **Cailleux Scale** This is a more lengthy method for determining particle 'roundness'. Take a pebble from the sample and measure its long axis in mm. Hold the pebble flat and lay the sharpest corner on the chart of semi-circular concentric rings to asses the radius of this corner (Figure 2.55). Use the formulae below to work out the sediment roundness.

$$R= (2r/a) \times 1000$$

Where R = roundness index. 'r' = radius (in mm) and 'a' = length of long axis (in mm).

e.g. r = 10,   a = 35

$$R = 2 \times 10/35 \times 1000 = 571$$

TOP TIPS

In essence a low 'R' value indicates a stone which is angular, whereas a higher value indicates more rounding. Although it is theoretically possible to get an R value of up to 1000, in practice the results are likely to be much lower, often less than 400 even for sediments which appear to be rounded.

▶ **Zingg Shape** You can use a Zingg diagram/graph (Figure 2.57) to determine sediment shape. Simply measure the length of the stone's axes a,b and c in mm and then calculate the ratios as indicated on the graph. Then plot the data on the graph. Note down any clustering or grouping of sediments in any of the four zones indicated. Use this as a tool to compare different sediment sites – it can further help with your analysis.

TOP TIPS

Stone or pebble orientation will have two values, 180° apart, e.g. 20° or 200°, so it does not really matter in which direction the compass is pointed as long as the base-plate remains parallel with the stone long axis (Figure 2.58)

nature and direction of the depositional process. This might help in the determination of flow directions in relic ice or fluvial landscapes.

The orientation of sediments should be measured with a compass along the long axis and the results recorded in degrees on a suitable booking or recording sheet. If the sediment is inaccessible and its long axis cannot be determined in-situ, carefully remove thc particle and reposition it in the same long axis orientation on the ground surface.

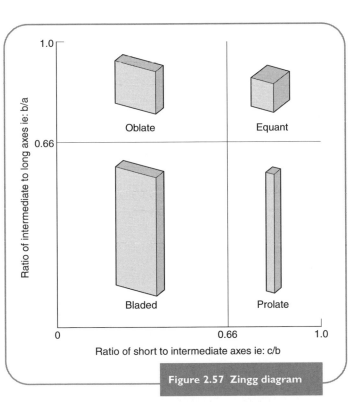

Figure 2.57 Zingg diagram

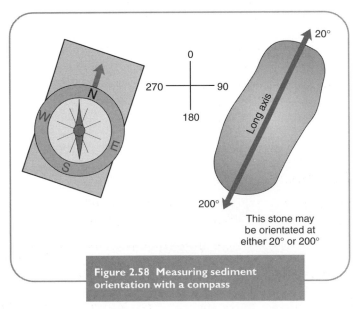

This stone may be orientated at either 20° or 200°

Figure 2.58 Measuring sediment orientation with a compass

Sediment orientation

The term 'fabric analysis' is sometimes used to refer to the three-dimensional disposition of the sediment. The positioning of a particle (lodged in a matrix) can allow us to make inferences about the

Orientation data are best plotted on a rose or star diagram (Figure 2.59). It is also important to calculate the mean, mode and standard deviation of all the datasets so that more reliable (and quantitative) comparisons can be made between sites or locations.

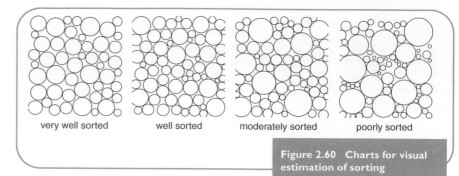

very well sorted   well sorted   moderately sorted   poorly sorted

Figure 2.60   Charts for visual estimation of sorting

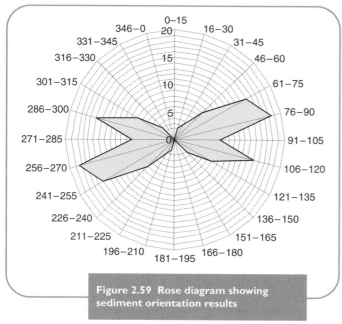

Figure 2.59 Rose diagram showing sediment orientation results

Degree of sediment sorting

It may be feasible to make a simple visual estimation of the degree of sorting of a particular deposit, i.e. how mixed the particle sizes within a sediment are. Use Figure 2.60 as a tool. The degree of sorting will provide clues on the nature and mode of deposition of a sediment. If a sediment is well sorted it may indicate the presence of water, whereas poorly sorted sediments are usually associated with direct ice contact, where material is quickly dumped as the ice melts.

How to interpret sediment results

Interpretation of your results should be the most engaging part of your study. Imagine that you are unlocking a secret door to the past, but to locate the key you need to get clues from the location, aspect,

	Glacial deposits			Table 2.9
Type of Deposit	**Process**	**Landform and likely location**	**Orientation**	**Shape**
Till	Results from (basal) melting, near or at the bottom of an ice mass.	Often found in sheets or displaying hummocky topography.	Unsorted, possibly compacted. Largely not orientated.	Mixture of sediment sizes and shapes. Tends to be dominated by angular material.
Soliflucted head	Mass movement of material, occurring in periglacial conditions.	Found on the side of valleys, sometimes in cusps or lobes, identified as large bulges.	Unsorted, long axes aligned parallel to direction of flow – always down-slope.	Angular clasts, may incorporate fine, wind blown material (loess).
Fluvio-glacial	Deposited from pro-glacial streams near ice or from streams flowing inside the glacier.	Typically pebbly/gravel alluvial outwash fans, braided in nature. Could also be eskers.	Relatively well sorted, tends to be orientated down stream and down valley but with some variation.	Coarse material dominant, mainly sub-rounded to sub-angular. Expect a mixture of other angular material.
Fluvial	Post-glacial stream	Can be meandering post-glacial stream. May show evidence of meander migration.	Well sorted, strong orientation in a down valley and down stream direction.	Mostly well rounded deposits.

orientation, shape etc. of your sediment. This process of critical examination and elimination is vital if meaningful and reliable conclusions are to be drawn. Table 2.9 provides a basic summary of the types of deposit that might be encountered during a field exercise in a glacial environment.

In the next section, our 'Ice Detective' will take you through their interpretation of field results. The following figures have been used by way of evidence (these are found within this chapter):

▶ Figure 2.51 the morphological map of the site,

▶ Figure 2.53 the field sketch of the valley.

What the Ice Detective might think......

Sediment Site A This site is located adjacent to the present river and is part of a river cliff exposure. The sediments at this site show a strong preferred orientation NNE-SSW, although a certain amount of bi-modal trend is apparent in the NW-SE axis. The Powers data indicates variability in sediment shape, but has a 'Rounded' modal class. Cailleux data which has been collected from this location also indicates a greater range of sediment shapes in comparison with Site B – a standard deviation of 178 at this site compared with 128 for Site B. Evidence, therefore, points to this deposit being fluvio-glacial in origin. The bi-modal trend from the orientation data can be

linked to a dynamic braided system on the valley floor during deglaciation, with the river regularly changing direction and course.

Sediment Site B The position of this site is on a small rise, about 20m above a small tributary. The Powers data indicates an 'Angular' mode, and there are relatively few sub-rounded or rounded stones. The Callieux data collected for Site B displays a mode of 101–150, and a mean value of 185 (compared to 289 at Site A). The preferred orientation of the sediments is in an east-west direction (remember the valley runs north-south). This is therefore in a down-slope direction. All the evidence points to the sediment being part of a soliflucted, or head deposit.

WEATHERING STUDIES

Within some upland environments, it is sometimes possible to compare rates of weathering between different areas which might have contrasting aspects or altitudes. Although it is almost impossible to measure *actual* rates of physical or chemical weathering without technical equipment, subjective assessments can be made based on simple field evidence. Use the scale below, a modified version of Rahn's Index to assist in your classification of weathered material.

CLASS	DESCRIPTION
1	UNWEATHERED – Looks fresh and almost new
2	SLIGHTLY WEATHERED – Evidence of discolouration, faint rounding of corners
3	MODERATELY WEATHERED – Surface not smooth, mild pitting
4	BADLY WEATHERED – Rough appearance, discoloured surface layer
5	EXTREMELY WEATHERED – Surface peeling, pitting and layering

TOP TIPS

The modified Rahn Index can also be used to approximate rates of weathering on gravestones, within different quarries or contrasting coastal environments.

Hypotheses and key questions can be developed relating rates of weathering to variables such as rock type, age of structure, aspect and degree of exposure. Sandstone, slate and limestone, for example, tend to weather rapidly, whilst fine grained igneous rocks like basalt have a much longer surface life expectancy. Use geology maps to determine rock types within your field area, and large scale OS maps to select suitable sites with contrasting aspects and exposures.

MOUNTAIN PROJECTS

This is a rather mixed bag of ideas, but it refers to projects that could be undertaken if you are fortunate enough to visit an interesting area for your field trip, such as the Alps or Iceland. Below are some sugesstions for projects, which could be basic 'set' fieldwork days.

Tourism

▶ Understanding the factors and reasons for the growth of ski tourism in Alpine areas.

▶ How has ski tourism both positively and negatively affected Alpine areas?

▶ A study of the major phases of spatial growth of winter sport stations/areas.

Economy and tradition

▶ The declining rural population and issues of abandonment.

▶ How and why can access and services be improved?

▶ Changes in the traditional economy, e.g. loss of pastoralism and replacement by tourism.

▶ The development of HEP, metallurgical and electro-chemical industries – a study of the changes and effects on river and air quality.

Physical projects

▶ Monitoring the discharge from a meltwater stream. How much does it vary on a diurnal or weekly basis? What type and size of load is it carrying? How does this vary?

▶ Mapping the distribution, type and size of moraine deposits in an area.

▶ A study examining the distribution of crevases in a glaciated area, and relating this to gradients.

▶ Evaluating the effects of geology, relief and exposure in controlling the development of contrasting soil types.

▶ The effects of a ski resort/tourism on local water quality and stream invertebrates (upstream and downstream of the settlement).

These projects can be undertaken whilst visiting the area, but you need to ensure that you have good access to secondary data to support any initial field findings as it will be unlikely that you can revisit your field site.

ACTIVITY AND AMENITY SURVEYS

Many upland glaciated regions have become popular visitor attractions; the Lake District and Snowdonia are among two of the most well known examples. Within such landscapes there will often be areas where visitors concentrate, i.e. honeypot sites. The box below outlines a number of possible studies in such regions.

Possible study	Data collection technique (s)
Conduct user surveys to establish the level and frequency of usage of a honeypot site	Simple questionnaires can used to determine the sphere of influence for particular honeypot sites. If post-codes are obtained for individual users, then the perception of distance could be evaluated, i.e. comparison of actual/perceived travel time vs straight line distances. Consider whether site accessibility influences the type of user.
An evaluation of the quality of tourist amenities	Particular issues could be explored, e.g. car parking problems, noise pollution and overcrowding. Make a detailed site map and record car parking/overcrowding 'hotspots'. These could also be considered on a seasonal basis, by interviewing wardens/site managers. The study can be developed by suggesting solutions to daily/weekly/monthly overcrowding issues (which may need different solutions or management strategies).
Recreational surveys	Record the number of users of a site over a one hour period, classifying them by age, gender and user type (e.g. dog walker). Do different types of users, i.e. walkers, bikers, sight-seers, use the resource in different ways? They might favour particular areas in preference to other 'zones', whilst some parts of the resource might be used by the majority of visitors, e.g. toilets. If a detailed base map is made, patterns, type, frequency of movements etc. can be plotted using flow lines and pictograms. Impacts on landscape are well documented.

WEBSITES

www.virtualmontana.org/virtualmontana – The Virtual Montana website contains mountain links and examples of students work. Good focus on glaciation

daac.gsfc.nasa.gov/DAAC_DOCS/geomorphology/ GEO_9_CHAPTER_9.HTML – some quality facts and figures about world glaciation

www.zephryus.demon.co.uk/geography/resources/ revision/ice.html – good glaciation glossary can be found at this location

www.zephryus.demon.co.uk/geography/resources/ glaciers/tour.html – lots of photos of a virtual glacial 'tour'

vulcan.wr.usgs.gov/Glossary/Glaciers/framework. html – big glaciation site.

WEATHER

Figure 2.61 Weather station instruments

STEVENSON SCREEN

ANEMOMETER

SOIL THERMOMETER

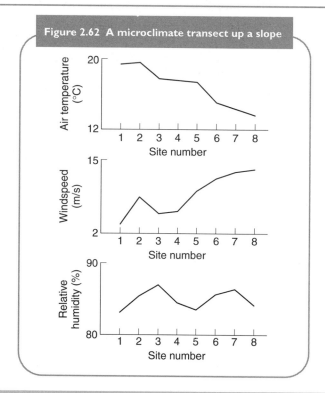
Figure 2.62 A microclimate transect up a slope

Weather is a description of the state of the atmosphere at a given time. It describes the day-to-day changes in temperature, precipitation, cloud cover, sunshine, wind, pressure etc. Generally there are two different types of weather study:

Weather This is the study of weather at one site, perhaps involving observations made over a day, week or even longer. Frequently variables such as wind speed and direction, temperature, hours of sunshine, cloud cover and humidity are recorded (Figure 2.61).

Microclimate This is the smallest scale of study in climatology and it focuses on the relationship between the physical characteristics of the Earth's surface and the conditions in the shallow layer of the atmosphere immediately above it. In other words it is the study of climate near the ground (more specifically the climate below the level of instruments used in meteorological observation) (Figure 2.62).

WEATHER PROJECTS

The range of projects available includes the following.

Weather Station	Microclimate
Relationship between air pressure, precipitation and wind velocity.	Changes in humidity and wind speed with increasing distance away from a lake or pond.
Contrasting stations: urban vs rural, sheltered vs exposed, low altitude vs high altitude, north vs south aspect.	Comparison of microclimates in coniferous and deciduous woodlands.
The study of a cold front as it passes over a weather station; changes in cloud type/cover, wind speed and direction, temperature and pressure.	Different microclimates on the edge and interior of a woodland.
The study of a warm front as it passes over a weather station; changes in cloud type/cover, wind speed and direction, temperature and pressure.	Changes in air temperature and wind speed from the bottom to top of a slope.
Comparison of soil and air temperature variation over one week.	Measurement of air temperature differences between town and countryside.
Home made vs standard equipment: a discussion and review of the strengths and limitations of each type of instrument.	The diurnal changes in air temperature for a named British city.

The following techniques are discussed in this chapter.

Technique	Weather station (W) or microclimate (M)
Temperature: air, ground and soil	Must use specialist equipment, suitable for both W and M.
Precipitation	Can use home made equipment, suitable for both W and M.
Pressure	Needs barometer or equivalent, W and M.
Relative humidity	Needs specialist equipment, W and M.
Wind speed and direction	Can use home made equipment eg hydrometer, but scientific equipment better, W and M.
Cloud cover and type	Needs key e.g. Field Studies Council. Most suitable for W.
Visibility	Needs base map. W only.
Radiation and sunshine	Needs specialist equipment, W and M.

Temperature

Temperature is measured with a thermometer:

▶ Digital thermometers give precise readings to 0.1°C, but can be unreliable if incorrectly calibrated, if they get wet or the batteries run low.

▶ Analogue thermometers (mercury or alcohol) are made of glass so there is risk of breakage, but they are reliable and should be accurate to at least 0.5°C.

Temperature observations can be made in three main positions/situations:

1. Air Atmospheric temperature is recorded by a thermometer placed inside a Stevenson Screen. Alternatively, carefully fix a thermometer to a metre rule and insert this into the ground.

2. Ground These temperatures are more extreme than air temperatures. Support a thermometer horizontally on small blocks 1–2 cm above the ground, or rest the temperature probe at this height. Do not let the probe come into direct contact with the ground surface.

3. Soil Dig a small hole in the ground, 20–50 cm deep. Lower the thermometer into the hole, loosely in-fill and leave for 3 minutes. Remove and quickly read the temperature. Soil probes can be pushed straight into the soil.

Using a maximum and minimum thermometer (Figure 2.63) the daily temperature range can be measured. This is the difference between the hottest and coldest air temperatures each day – small markers on the thermometer indicate maximum and minimum extremes.

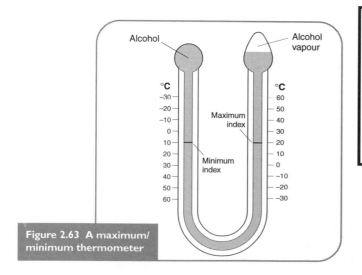

Figure 2.63 A maximum/minimum thermometer

Precipitation

Precipitation is any liquid or solid that condenses from water vapour, i.e. rain, drizzle, snow, sleet and hail. Precipitation is measured as liquid volume, normally using a rain gauge. This can be home made (a funnel and jar) (Figure 2.64) or can be purchased cheaply. If you are going to make several rain gauges ensure that they all have the same diameter funnel so that it is a fair test (the British standard gauge is 5 inches).

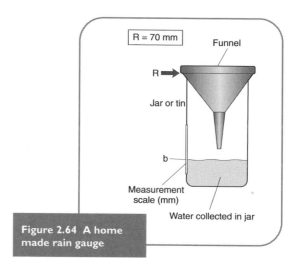

Figure 2.64 A home made rain gauge

Site rain gauges carefully – buildings and trees will affect the results. Gauges should be sited 10 cm above the ground surface to reduce rain-splash effects. Leave out for 24 hours and transfer the water collected (if any) to a measuring cylinder, then, calculate the amount of precipitation in mm as shown below.

$$\frac{1000 \times \text{volume of rain in mm or cc}}{3.14 \times R \times R}$$

Where R is the radius of the funnel in mm

Example:

$$\frac{1000 \times 6}{3.14 \times 70 \times 70} = \frac{6000}{15386} = 0.39 \text{ mm in 24hr.}$$

Pressure

The weight of air in the Earth's atmosphere exerts a force known as pressure. High pressures can reach as much as 1040 mb (millibars), while low pressures can go down to 960 mb. Air pressure is measured with a barometer. There are mercury barometers available, but it is more likely that you will be able to borrow an aneroid barometer. These are straightforward to use, relatively accurate and fairly reliable if kept indoors (air pressure is the same inside as outside).

Relative humidity

Humidity is the amount of water vapour present in the air. Relative humidity (RH) is measured as a percentage and is the amount of moisture in the air compared to the amount needed to saturate the air. If the air is saturated then the RH = 100% .

Relative humidity is based on the differential readings between 'wet' and 'dry' bulb thermometers. If the readings on both thermometers are the same, then the air is saturated. If there is a small difference between the two, the humidity is high; a large difference between the two indicates lower humidity.

Relative humidity can be measured using two instruments:

▶ **Wet and dry bulb hygrometer** These are usually found fixed within a Stevenson Screen and so are not suitable for microclimate measurements.

▶ **Whirling psychrometer** The device is whirled above the head (rather like a football rattle) for 1 minute and readings of the two thermometers are taken.

Wind speed and direction

There are a number of ways to record wind speed:

▶ **Anemometer** This works in near still conditions, is effective and reliable, but cost may be prohibitive at around £100–150.

▶ **Ventimeter** This is a cheaper instrument at around £20–30, reasonably accurate but does not work in low wind conditions (Figure 2.65).

▶ **Beaufort Scale** This requires no equipment, it is based on a qualitative assessment based on subjective descriptions, *e.g. smoke rising vertically = 0* (Table 2.10).

Wind speed is the most variable of all weather phenomena, and can change from second to second. Observers should take recordings over a period of several minutes to obtain a reliable mean.

Wind direction can be determined from a weather vane (e.g. on a local church) or from a wind sock (found on air fields).

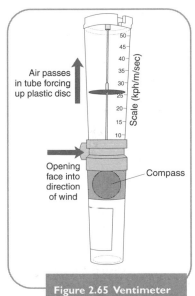

Figure 2.65 Ventimeter

			Beaufort Scale (wind speed)	**Table 2.10**
0	◎	Calm	Smoke rises vertically
1		Light air movement	Smoke drifts in the wind
2		Light breeze	Wind felt on face
3		Gentle breeze	Leaves and small twigs move
4		Moderate breeze	Loose paper blown about
5		Fresh breeze	Small trees start to sway
6		Strong breeze	Larger branches move
7		Near gale	Difficult to walk into wind. Branches may break off
8		Gale	Possible structural damage – tiles may be blown off roofs
9		Strong gale	Structural damage more likely, cars may be rocked around
10		Storm	Trees uprooted, doors blown open
11		Violent storm	Very rare, widespread damage
12		Hurricane	Extreme weather event – unlikely to occur in Britain

Often it is easier and more practical to assess wind direction yourself. Throw some grass or a tissue in the air, then use a compass to gauge the wind direction.

Cloud cover and type

Clouds are accumulations of water droplets or ice crystals – their type and coverage indicate the moisture content and temperature of the air. Cloud cover is estimated in 'oktas' or eighths of the sky. It is difficult to assess cloud cover as the whole of the sky must be assessed at one time. Table 2.11 shows the symbols used to indicate the amount of cloud cover.

Figure 2.66 shows the major cloud types and the weather conditions associated with each.

▶ **Cirrus** Fibrous or hair like

▶ **Cumulus** A heap or pile

▶ **Stratus** A horizontal sheet/layer

▶ **Nimbus** Rain bearing.

Visibility

In clear weather, make a sketch map of easily recognisable landmarks at different distances from the observation site. Using a large scale, 1:25,000 OS map, measure the straight line distances with a ruler or

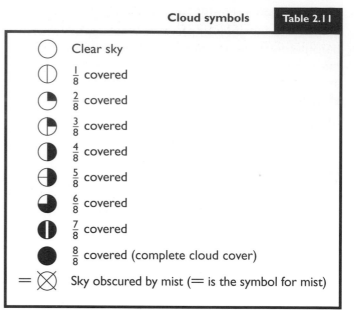

Cloud symbols   Table 2.11

○	Clear sky
	$\frac{1}{8}$ covered
	$\frac{2}{8}$ covered
	$\frac{3}{8}$ covered
	$\frac{4}{8}$ covered
	$\frac{5}{8}$ covered
	$\frac{6}{8}$ covered
	$\frac{7}{8}$ covered
●	$\frac{8}{8}$ covered (complete cloud cover)
⊗	Sky obscured by mist (= is the symbol for mist)

TOP TIPS

A good observation point is one which is not obscured by buildings or trees, where there is a 360° vista, preferably on raised ground.

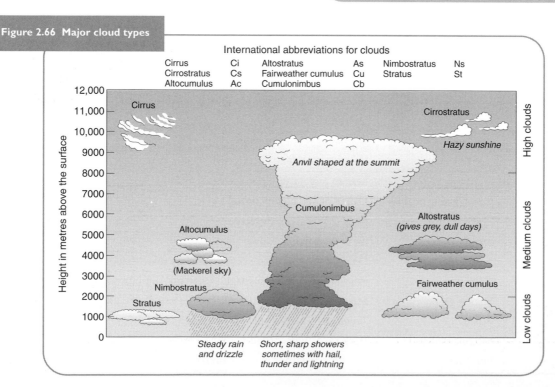

Figure 2.66 Major cloud types

International abbreviations for clouds

Cirrus	Ci	Altostratus	As	Nimbostratus	Ns
Cirrostratus	Cs	Fairweather cumulus	Cu	Stratus	St
Altocumulus	Ac	Cumulonimbus	Cb		

piece of string. Number and rank the features according to distance. Visibility can be measured by the furthest point visible at the time of observation.

Radiation and sunshine

Although difficult to measure, the most scientific way of estimating radiation is with a light meter. Be careful with your methodology as the light meter will record reflected light as well as incident light.

On a fixed weather station, sunshine hours are usually recorded with a Campbell-Stokes sunshine recorder.

WEATHER STATION STUDIES

Monitoring the weather in one location requires careful and accurate monitoring. To ensure reliable readings, observations should be taken at the same time on each day (this is usually done at 9.00 GMT, although you could choose any time that is convenient to you). Figure 2.67 is an example of a weather log and Table 2.12 shows a booking sheet. Provide a brief site description at each location where measurements are taken as site features may

TOP TIPS   Measuring the weather

It is important to keep accurate and reliable readings of any aspects of the weather which form the basis for study. It is also important, once the data has been collected, to provide visual summaries in the form of graphs.

Figure 2.67 Summary weather log

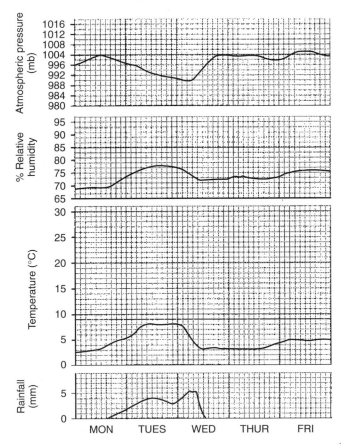

Summary Graphs of Weather Data – for the week ending...... December 20th 2002

Site of weather station: Birmingham

Altitude: 50 m

Grid reference: SK0996

Observation time (GMT): 9.00 am

Cloud types observed:

Cirrus, Altostratus, Nimbostratus, Cumulonimbus

Cloud cover in oktas

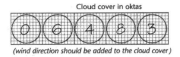

0 6 4 8 3

(wind direction should be added to the cloud cover)

| Weather observation sheet | | | | | | | | | | | | | | | Table 2.12 |

Date	Cloud cover in oktas	Cloud type	Temperature (remember to reset the thermometer when you are recording)			Relative humidity			Precipitation (empty gauges after recording)	Visibility distance and light intensity	Wind		Pressure (mb)	Remarks
			°C Max	°C Min	°C Mean	Dry bulb °C	Wet bulb °C	%			Mean direction	Mean speed		e.g. thunder, lightning, smoke, haze, gusts, wind chill
am														
pm														
am														
pm														

well influence your results, e.g. overhanging trees or tall buildings may affect wind speed, sunshine, temperature etc. Take photographs as evidence, and estimate the height of any relevant buildings, trees or walls which may have an effect.

Most weather projects, involve the observation and recording of local weather over time. Often you will need to relate local changes in atmospheric processes to the regional atmospheric circulation. For this reason it is important to obtain daily records from secondary sources in order to set the local weather changes in the regional context and develop explanations of changing weather patterns. These secondary sources can include MetFAX, newspapers, shipping forecasts, or weather data from websites (see Table 2.13).

Sources of weather data			Table 2.13
	Advantages	**Disadvantages**	
MetFAX	Quick and easy to get hold of and regularly updated. Can also get 'Weather Packs' (includes charts and satellite images).	Must set fax machine to 'poll' mode. Transmissions are charged at premium rates.	
Newspapers	Easy to get hold of and cheap. Readily interpreted.	Results are a day late. Often generalised by region. Could be coverage issues, depending on nature of paper.	
Shipping forecasts: BBC R4	Accurate and reliable forecasting. Can use atmospheric data to plot your own weather maps.	Can be difficult to interpret, also knowledge of shipping areas required (need a base map). Forecasts often at inconvenient times.	
Websites: www.met-office.gov.uk www.bbc.co.uk/weather	Up to date weather information, and weather data for recent months.	Speed of download my be an issue, together with 'on-line' cost. Can be difficult to navigate to find exactly what you want	

Possible ideas	Possible study
General observations	Collect relevant data and consider the following: ▶ Is there more rainfall with low barometric pressure? ▶ Is there a high wind velocity associated with low pressure? ▶ Does ground temperature vary more than air temperature?
The study of a depression as it passes over the weather site	Study a warm or cold front as it passes over. Note what changes there are in terms of cloud formation, wind speed and direction, barometric pressure and temperature. Make comparisons between your observed results and the classic textbook model. Use ICT to make analysis and comparisons more convenient, i.e. suggest relationships: rainfall v pressure over time. Back up your findings with reference to published satellite images and synoptic charts, or consider producing your own weather chart.
Comparing weather stations	Either set up you own mini-weather stations in contrasting localities or use secondary data, then make comparisons, e.g: ▶ Is the coast milder and wetter compared with inland? ▶ Is there a difference between rainfall on the west and east of slopes? ▶ Is there a difference between temperatures in cities and in the countryside? ▶ Is there an observable difference in temperatures and sunshine hours on northern slopes compared to those with a southern aspect? ▶ Comparison of wind speed frequencies for two contrasting sites.
Evaluating the reliability of 'home-made' weather instruments	Using the same siting areas, set up your own home-made instruments alongside an existing climatological/weather station. Which observations are most reliable and which measurements are least effective? Try to evaluate the strengths and limitations of your equipment and suggest design modifications of how you might improve them for future use.

TOP TIPS

Weather and air pollution

Air quality data can be gathered from secondary sources and compared with your own primary weather data, i.e. to test the relationship between air quality and weather.

Air quality data is available from a range of sources, but a good internet location is the DETR website:www.environment.detr.gov.uk/airqual/aqinfo.htm, or the Environment Agency: www.environment-agency.gov.uk The number and spread of locations available should allow a site reasonably close to home/school to be used.

Collect data on windspeed, direction and atmospheric pressure, and see whether there is any relationship between the weather conditions and levels of pollution, examining in particular the effect of 'dispersion' in terms of affecting air quality.

You could also use secondary data to explore variations in pollutants for different parts of the Britain and compile pollution maps, identifying pollution 'hotspots' under different weather conditions. Downloadable satellite maps are available from the Met Office and BBC websites.

MICROCLIMATE STUDIES

Spatial variations in the form and character of surfaces can create readily measurable differences in microclimate over very short distances. Whatever the local environment or area, there are usually many opportunities for geographical fieldwork. Field data collection can be based on one of the following:

▶ Short term field exercises which can be completed in a morning or afternoon.

▶ Repeated short term experiments under contrasting synoptic weather conditions.

▶ Longer term observations involving the installation of monitoring equipment in the field, such as rain gauges and a regular measurement routine.

Projects can either be almost entirely based on primary first hand data, or a combination of primary and secondary data.

Sampling procedures

Many microclimatological studies are based on the sampling of a transect, which, although labour intensive, usually generates the most reliable datasets. However, the spatial complexity of microclimates makes sampling a problem and considerable care should be taken in the selection of monitoring sites. There are two main sampling frameworks that can be used.

▶ **Spot measurements** These are taken at 2–6 carefully selected sites and are taken *simultaneously* by small groups. They provide datasets which can be compared using simple averages and ranges. A range of microclimatic variables, e.g. different altitudes, aspects, locations etc. can be obtained.

▶ **Transect** In this method climatic variables are measured at a series of points separated by a fixed distance (systematic) or by altitude (stratified). Usually 8–12 sites might be chosen to give a good coverage (and allow statistical testing).

Topoclimates

Local topography (the nature and shape of the land) generates localised microclimates, which are called topoclimates. Topoclimates are characterised by small scale variation over distances measured in terms of hundreds of metres. There are four primary topographic controls:

▶ aspect

▶ angle of slope

▶ degree of exposure

▶ altitude.

There are various types of topoclimate project that can be undertaken. Examples include:

▶ How wind speed, humidity and air temperature change from the bottom to the top of a slope (Figure 2.68).

▶ How microclimatic variables vary between an exposed and sheltered site.

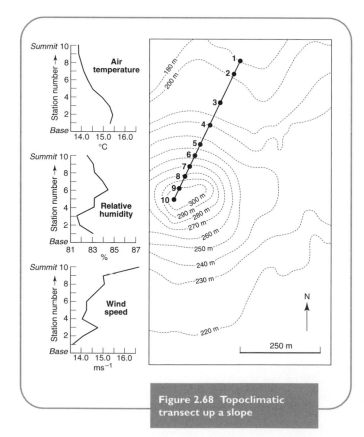

Figure 2.68 Topoclimatic transect up a slope

▶ What impact aspect has on temperature and number of sun-hours.

Time lapse error

As time elapses between setting out and arriving at the end point of your transect, any observed differences in climate may be due to the real effects of, for example, a difference in altitude, or they may be due in part to it being later in the day. In order to try and remove this time lapse error, the transect must be reversed and measurements taken at the same points on the way back to the starting point. The time taken for the return journey should be the same as for the outward journey. The results obtained from the two transects should then be averaged.

Topex method for assessing site exposure

From a site position, gradients of slopes (either up or down) can be measured to the nearest 1° along the

Topex values **Table 2.14**

	Angles (°)	
Compass direction	**Exposed site (hill summit)**	**Sheltered site (valley location)**
North (0/360)	0	+6
North East (45°)	+1	+8
East (90°)	+1	+5
South East (135°)	0	+7
South (180°)	−1	+3
South West (225°)	0	+3
West (270°)	+2	+2
North West (315°)	+1	+15
TOPEX VALUE	**+4**	**+49**

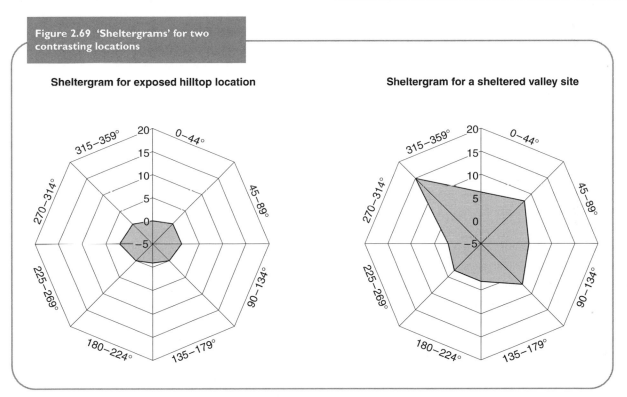

Figure 2.69 'Sheltergrams' for two contrasting locations

Sheltergram for exposed hilltop location

Sheltergram for a sheltered valley site

eight principle compass points (inter-cardinals) using a clinometer, i.e. N, NE, E, SE etc. Some example results are presented in Table 2.14 – the higher the 'topex value', the more sheltered the site. The measurements can be plotted in the form of a 'sheltergram', in which the central value is –5° (Figure 2.69) – this shows the degree of exposure visually.

School, garden and urban microclimates

The presence of buildings on the ground surface has a measurable effect on the local microclimate. Microclimates around buildings present a number of opportunities for local climate investigations which can be based on small groups of buildings such as the school complex, or a hamlet, or a large urban area. Spatially small areas for study, however, usually produce relatively subtle effects which will require

sensitive equipment. In such cases a more profitable line of enquiry may be the temporal variability due to the effect of buildings on airflow and energy balance. At a larger scale, towns and cities can generate very strong climatic contrasts with their rural surroundings, although often there will be logistical and safety considerations when undertaking such a project.

Sky View Angle (SVA)

This is a technique for mathematically determining the degree of exposure of a street based on the height and spacing of buildings. The Sky View Angle (A) is the angle made by the lines joining the tops of the buildings to a point in the middle of the street. In theory this could be measured easily with a clinometer by standing in the middle of the street – however it can also be simply derived from measurements taken from the relative safety of the pavement (Figure 2.70).

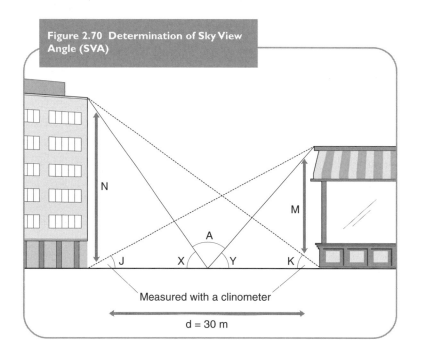

Figure 2.70 Determination of Sky View Angle (SVA)

Measured with a clinometer

d = 30 m

Study	How to do it	What you might find/graphical representation
Survey of seating and recreational use of school grounds – relating it to sunshine, temperature and wind speed.	Using a detailed large scale map of the school grounds select a number of sampling points based around the areas of recreational use. A pilot survey would help identify the popularity of such locations. Measure ground and air temperatures, together with wind speed and hours of sunshine.	People tend to sit in the warmest and most sheltered locations. Produce an annotated map to display your results showing the degree of exposure for each location. It may be possible to produce a simple school grounds management plan to help improve the seating and recreational use of the area.
Comparison of microclimate data for a school sports field and the school/college campus.	Draw a base map of the campus, selecting 10–15 sites randomly. Obtain reliable mean values for sunshine, temperature and wind speed/direction. Compare this data with 10–15 sites from an open area such as a school playing field.	Expect the open area to have higher average air flow values, but probably greater peaks around the buildings of the campus due to the channelling and 'venturi' effect. Plot data as graphs.
Exploring the urban heat island.	Conduct a pilot survey along a transect from the city centre to the outskirts. This can be achieved using the air temperature gauge on a modern car to confirm there are sufficient temperature differences. Then carry out a more detailed systematic survey using more accurate thermometers and a number of sampling points.	Expect there to be an observable temperature difference between the 'heat island' and the rural fringe. If you conduct a number of surveys radiating out from a city centre then try to produce an isotherm map using the data collected. (As a general rule, at least 30 measurements should be undertaken to produce such a map.)
Street exposure (Sky View Angle SVA) and urban temperatures.	For details of this technique see page 96. You will need to obtain a detailed street map to select a number of sampling sites.	The technique assumes that the taller the buildings, the greater the retention of heat at ground/street level. Plot the data using suitable graphs and base maps.
Comparing humidity inside and outside buildings.	Use a whirling psychrometer to measure humidity at a number of locations both inside and outside (perhaps totalling 20–30). Try to obtain a reliable mean for each environment and see whether there is a difference between morning and evening, or day and night. Alternatively you could compare winter/summer and spring/autumn.	The lower relative humidity indoors can be attributed to fewer sources of water vapour. There may also be a link between number of people and humidity.
Fog and snow surveys.	A relatively simple survey is to identify areas within a park, village or small town where snow tends to accumulate or where fog is prevalent. Use a detailed base map and measure snow depth with a metre stick, or fog using a simple sliding scale based on visibility. Select a number of different sites and compare results.	Try to examine how and why there are differences in the accumulation of snow and fog in an area. In particular, try to explain what effect buildings have on snow accumulation or the development of fog patches. It might be possible to test whether there is a relationship between air/ground temperature and the formation of fog.

The height of the buildings = M and N and the street width = d; the angles measured to the top of the buildings from the opposite sides of the street = J and K. Note J and K are known values, recorded in the field with a clinometer. In Figure 2.70 d = 30m, J = 35°, and K = 50°. There are three steps needed to complete the calculation and derive the SVA (A):

Use the SKV to look at exposure of different streets and see how this relates to diurnal temperature fluctuations.

Step (1) Calculate the heights of buildings M and N
M = d.tanJ, so M = 30 × 0.70 = **21.00m**
N = d.tanK, so N = 30 × 1.19 = **35.75m**

Step (2) Calculate angles X and Y
Y = tan^{-1} (2.tanJ), so, Y = tan-$^{-1}$ × 1.40 = **54.46°**
X = tan^{-1} (2.tanK), so, X = tan^{-1} × 2.38 = **67.21°**

Step (3) Work out angle A
A = 180 − (X +Y), so A = 180 − (54.46 + 67.21)
A = **58.33°**

Woodland microclimates

Investigations on woodland microclimates are most likely to be based on measurements taken along a transect or on comparisons between sites within and outside a wood. The variable nature of tree canopy means it is essential to take measurements at a group of sites rather than at a single site with the wood, so that reliable averages can be determined. There are a number of possible studies to consider.

1. **Solar radiation** The most straightforward investigations use comparisons of light intensity inside and outside a wood. A light meter, which gives a numerical indication of light intensity can be used to generate the two datasets. The means and standard deviations can be calculated and simple statistical tests carried out. An additional idea is to take measurements under a number of different tree canopies to see which canopy type delivers the greatest and the least solar radiation. Transmission coefficients can also be determined for a particular woodland type, e.g. oak.

Use a fixed grid system of sampling to calculate T_c values for a woodland – these can then be plotted up in the form of an isopleth map (Figure 2.71). Note a higher T_c value indicates less shade effects, hence a higher amount of light passing through the canopy to the woodland floor. It may be possible to establish a relationship between the transmission coefficient and the type or diversity of plants found on the woodland floor.

Transmission coefficient:

$T_c = L_I/L_o$

Where L_I = light level inside the wood and,
L_o = light level outside the wood

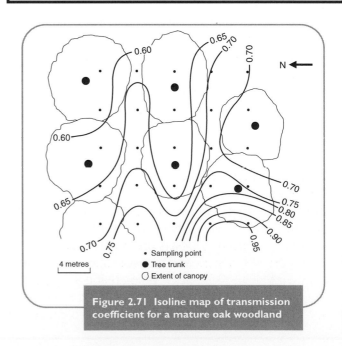

Figure 2.71 Isoline map of transmission coefficient for a mature oak woodland

2. Temperature Comparisons can be made between the air and ground temperatures inside and outside a wood. A comparison of soil temperatures taken under deciduous and coniferous canopies, for example, will show a marked effect on maximum temperatures. Alternatively, measure the hourly temperature of two contrasting sites over a 24 hour period. Example results are shown in Figure 2.72.

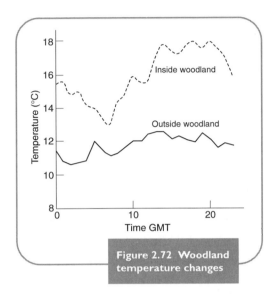

Figure 2.72 Woodland temperature changes

3. Airflow In the centre of a dense woodland there will be a predominance of slack or calm wind conditions. These decreases in wind speed can be measured with hand held anemometers (although they are insensitive to low wind speeds). Figure 2.73 shows measurements of airflow taken along a transect into a deciduous forest. Repeated measurements along the same transect will allow the calculation of the standard deviation for each point. Alternatively data can be collected to construct an isovel map of the wood, showing the spatial variations in wind speed over the area.

4. Humidity The moisture content within a wood is determined by the amount of mixing of water vapour by air. This can be measured with a whirling psychrometer along a transect, but the results are usually very variable and difficult to interpret. A more reliable outcome can be achieved by (statistically) comparing two sets of relative humidity values from within and outside a woodland, where there should be more measurable differences.

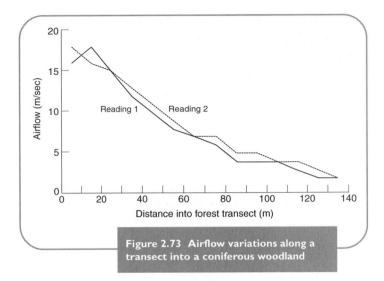

Figure 2.73 Airflow variations along a transect into a coniferous woodland

5. Precipitation Rainfall which is intercepted by a tree canopy on its passage to the ground surface may continue its journey by trickling down the branches and trunks of trees as stemflow, or may drip from the leaves as throughfall. When stemflow and throughfall are added together the total is usually less than the rainfall arriving at the top of the tree canopy. This deficit – the interception loss – is the result of evaporation of water from the leaves.

Interception = rainfall – stemflow – throughfall

It is possible (although ambitious) to measure some of these components of the water balance of a woodland canopy. Stemflow can be estimated by tightly wrapping a length of small diameter flexible tubing around the trunk of a tree and sealing it with mastic, or using plasticine to make small troughs. A tube connected to the troughs/pipe can be used to drain the collected water into a measuring cylinder (Figure 2.74). Rainfall in the woodland can be monitored with 20–30 rain gauges spaced in a random or grid pattern over the floor of the wood. Rainfall at the top of the canopy can be determined by placing a single rain gauge at least 100m away from the wood. Experiments can be set up to make comparisons between different types of wood, or the same woodland under different rainfall durations (the assumption being the longer the duration, the lower the interception loss).

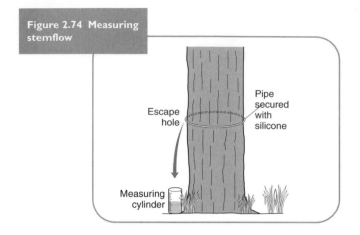

Figure 2.74 Measuring stemflow

Escape hole

Pipe secured with silicone

Measuring cylinder

Water microclimates

The critical factor determining the impact of a body of water on microclimate relates to its size or spatial scale. As a general rule, bodies of water less than 1 km^2 are unlikely to create microclimate effects which are readily detectable; however there are likely to be considerable microclimatic effects induced by a large body of water such as the sea. Project opportunities can be based on the analysis of climatological data at both regional and local levels, or the collection of field observations within limited distances from the water's edge.

1. **Sunshine** In many coastal locations the average hours of sunshine may be markedly higher than inland. Comparisons of sunshine hours can be made between coastal and inland sites. Collect data from daily newspapers or research climate data on the internet. The Meteorological Office also publishes a monthly weather report which would be useful.

2. **Temperature** The difference in thermal properties between water bodies and the adjacent land surface results in slower heating and cooling rates in water bodies which can be detected over considerable distances. At a local scale, regular measurements of air temperature over the course of a day at a coastal site and an inland site more than 30 km away should reveal a greater maximum and lower minimum at the inland site. An alternative approach on a micro-scale is to conduct a beach or inter-tidal transect. Across relatively wide beaches, in excess of 200 m between mean low water (MLW) to mean high water (MHW), a marked gradient of surface and sub-surface temperature develops on sunny days. Sand surface air temperatures should be recorded systematically along a transect at roughly ten places (Figure 2.75). Mark the observation points with painted stones or pegs. Ideally two or more transects should be carried out simultaneously and the results averaged. Sand temperature can be also be measured using a digital probe, sunk to the required depth.

As the beach dries out quickest nearest the upper part of the beach (heat is lost during evaporation), highest temperatures are expected near the top of the beach. Differences as high as 10°C can be recorded between contrasting sections of the beach.

3. **Airflow** The change in roughness as air moves from water to land means there is a rapid increase in friction between the surface and the atmosphere. On low-lying coastlines with little or no cliffs the rapid reduction in onshore wind speeds can be detected with hand held anemometers. It is also possible to demonstrate how forced uplift, over a storm beach or sand dune, for instance, produces accelerated wind speeds over the top of the feature. Using a transect and a number of systematic observation points it should be possible record the gradient of the feature and the wind speed at each point. These can then be plotted onto a cross-section diagram.

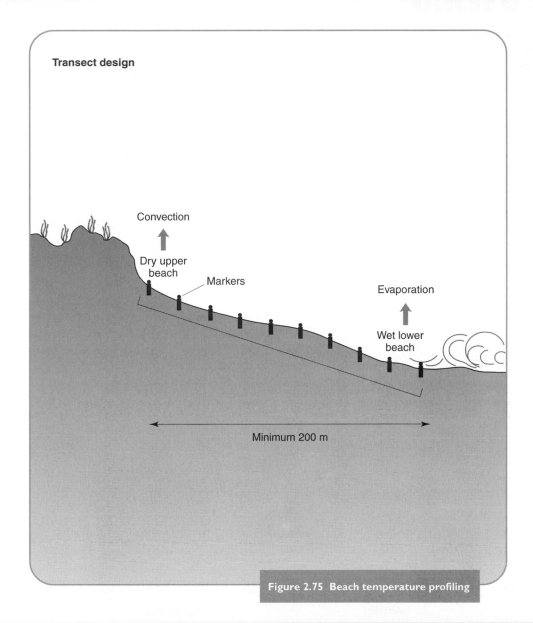

Transect design

Convection

Dry upper beach

Markers

Evaporation

Wet lower beach

Minimum 200 m

Figure 2.75 Beach temperature profiling

WEBSITES
Secondary data sources

www.esci.keele.ac.uk/weather – website detailing the current weather conditions at Keele University, Staffordshire. Also supplies monthly summaries.

www.met-office.gov.uk – the UK Meteorological Office website.

www.bbc.co.uk/weather – lots of weather information, satellite maps, etc. Also a useful weather glossary.

www.met.rdg.ac.uk/~brugge – contains download-able datasets for a range of climatological stations.

www.royal-met-soc.org.uk – The Royal Meteorological Society's home page. You can request free leaflets from here! Lots of other weather facts.

There are a number of microclimate articles featured in Teaching Geography, published by the Geographical Association, Sheffield. eg, 'Studies in microclimatology', Vol. 22, no. 4, October, 1997. See also 'Urban heat islands', Geographical Review, vol. 2, no. 3, January, 1998.

ECOSYSTEMS

Figure 2.76 Examples of ecosystems

SANDUNES

SOIL

ROCKY SHORE

MOORLAND

PARK

GARDEN

HYDROSERE

An ecosystem is a natural unit in which the life cycles of plants, animals and other organisms are linked to each other and to the non-living environment to form a natural system.

Wherever you go – town or countryside – there are pockets and islands of vegetation. There are many examples of ecosystems – woodlands, wetlands, moorlands, roadside verges, fields, gardens and parks (Figure 2.76). Given this variety of habitats and the issues which relate to them, there are many opportunities for studying ecosystems,

for example investigating how and why soil type and quality may be related to altitude and plant productivity. You might also choose to investigate what impact humans have on ecosystems, or consider your own alternative strategies for the management of an area. Some elements of ecosystem conservation and exploitation, for example, might also provide linkage to the more politically fashionable concepts of sustainability, stewardship, sustainable development and Local Agenda (LA) 21.

INVESTIGATING SOILS

Soil is an extremely complex medium. It is one of the key elements of the ecosystem, providing nutrients for plants as well as a medium for them to grow in. If soil is removed by the mismanagement of the landscape (e.g. deforestation) or by catastrophic events (e.g. high winds and floods), it can result in dramatic losses in vegetation and wildlife habitats.

Soil type, quality and depth can vary immensely within even a very small geographical area (see Figure 2.78). Yet all soils (anywhere in the world) are made of five basic 'ingredients':

▶ air
▶ water
▶ mineral matter
▶ rotten organic material
▶ living organisms.

Pedologists (soil scientists) also recognise five factors which control and limit soil type:

▶ geology/parent material
▶ climate
▶ relief/topography
▶ living organisms (including the impact of people)
▶ time.

The following are ideas for projects involving soil analysis.

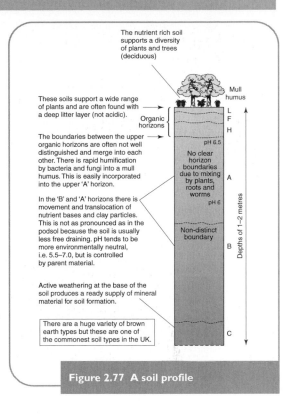

The nutrient rich soil supports a diversity of plants and trees (deciduous)

These soils support a wide range of plants and are often found with a deep litter layer (not acidic).

The boundaries between the upper organic horizons are often not well distinguished and merge into each other. There is rapid humification by bacteria and fungi into a mull humus. This is easily incorporated into the upper 'A' horizon.

In the 'B' and 'A' horizons there is movement and translocation of nutrient bases and clay particles. This is not as pronounced as in the podsol because the soil is usually less free draining. pH tends to be more environmentally neutral, i.e. 5.5–7.0, but is controlled by parent material.

Active weathering at the base of the soil produces a ready supply of mineral material for soil formation.

There are a huge variety of brown earth types but these are one of the commonest soil types in the UK.

Mull humus

Organic horizons: L F H

pH 6.5

No clear horizon boundaries due to mixing by plants, roots and worms

pH 6

Non-distinct boundary

Depths of 1–2 metres

A

B

C

Figure 2.77 A soil profile

Possible study	Data collection and presentation ideas
1. An investigation of how soil pH, texture, colour and depth changes from the top to the bottom of a slope – The classic soil 'catena' study.	▶ Use OS maps to find a suitable slope/area. Try to keep the distance short, i.e. up to 150m. Look for possible changes in geology which will also control soil type. ▶ Measure soil variables along either a continuous or interrupted systematic/stratified transect, i.e. at 10–20 locations. Use an auger to extract samples to determine soil depth, pH, colour etc. ▶ Also record gradient at suitable intervals with a clinometer. ▶ Consider collecting data on changes in vegetation type/structure with distance along the transect. ▶ Plot results using a range of cartographic techniques, such as pie charts of soil texture, pH, proportional depth bars and slope profiles, and graphs to make interpretation easier.
2. Changes in soil horizons and humic content associated with different rock types.	▶ Using a geology map choose contrasting locations (but try to ensure that other factors such as slope and aspect are constant). ▶ Dig soil pits (3–5) and measure, pH, colour, texture etc. of each horizon. ▶ Bag soil samples to determine chemical content and amount of organic matter. Always take the samples from a set depth each time to ensure fair testing. ▶ Plot annotated soil profiles for each location (Figure 2.77), and also consider developing your own basic 'soil map' for the area.

3. A comparison of garden/school/park/woodland soils linked to controlling environmental factors.	▶ Decide on suitable sites and a sampling strategy, based on the differences you are trying to observe. ▶ Design a recording sheet (Table 2.17) and identify environmental factors which are likely to be significant in controlling soil, e.g. slope, aspect, vegetation type (which should be included on the sheet). ▶ Make a detailed site description at each location, including sketches and photographs. ▶ Calculate moisture and organic content. ▶ Plot soil profiles (Figure 2.77) on a detailed, annotated base map. Also plot moisture and organic content as bar graphs or similar.
4. The impact of compaction on soil drainage (either by machinery or people).	▶ Choose a number of contrasting locations which will show different degrees of soil compaction. ▶ Devise a suitable scale for visual assessment of soil compaction and identify evidence of compaction, e.g. tyre prints. ▶ Calculate soil moisture content in the laboratory. ▶ Draw isolines for infiltration, bulk density and compaction on a suitably scaled base map.
5. The influences of different land uses and/or different agricultural practices on soil texture and soil chemistry, e.g. coniferous trees of different ages.	▶ Initially start by producing a detailed land use map of the study area. ▶ Pick a number of sites, perhaps 5–10 for carrying out simple surface tests – texture, pH etc. ▶ If the site is subject to different agricultural practices then the landowner/farmer should have records that can be obtained. ▶ The study can be extended by carrying out detailed soil texture analysis, i.e. sieving, bulk density etc.
6. Development of a farm management plan to reduce soil erosion risk in critical areas.	▶ Do an initial site survey to establish which area(s) show signs of most erosion, i.e. bare surfaces, gullying etc. Keep the area you intend to survey small, perhaps 1–3 fields. ▶ Obtain a large scale and detailed map of the survey area. ▶ Construct a soil risk map (Figure 2.85). You will need a compass to determine aspect and direction, and a clinometer to establish slope angles. Take plenty of photographs to record evidence. ▶ Prepare a management plan to offer solutions to minimise soil erosion 'hotspots'.

Soil survey techniques

Before starting a soils investigation, some thought to planning should be given.

▶ Having decided on some initial aims and objectives, research the soils topic you intend to pursue. Bear in mind that soils work is potentially complex, so gathering background information is critical to support any arguments developed later in your investigation.

▶ Consult maps, such as OS, geology and land use in order to find suitable locations which are accessible and safe. Also find out whether there is any other secondary data available such as from the local studies library or selected internet sources. Past projects may also provide useful secondary information.

▶ Decide on the likely equipment needed, both in the field and for any laboratory analysis.

▶ Determine how you will present your soil data once it has been collected. An example of a labelled soil profile is given in Figure 2.77.

Use Table 2.15 to help you to decide the scope and scale of a possible soils investigation.

Depending on the scale of your project, soil analysis may be restricted to field assessments, laboratory work, or more likely both. Table 2.16 shows which soil survey techniques require field or laboratory analysis.

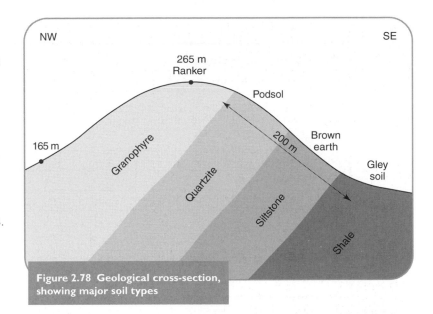

Figure 2.78 Geological cross-section, showing major soil types

Soil survey techniques | **Table 2.15**

Depth and 'level' of soil project	Your knowledge of soils and previous experience	Scale of project	Possible sampling ideas and resources
1 – Mostly basic [Data collected over a morning or afternoon]	Limited experience, limited knowledge base.	Small scale, within one area. Soils only form a (small) part of the whole project.	A few soil samples (3–5). Samples taken with trowel or similar, basic field analysis only from surface horizons. Little or no lab analysis. OS maps used.
2 – Moderate [Data collected over a ¾ or whole day]	Some understanding of soil process. Moderate knowledge base and reasonably confident.	Small/medium scale, perhaps involving limited comparative sites. Soil study is an integral component of the final report.	A number of samples taken (5–10) from different locations. Samples obtained with an auger, or a few samples collected from soil dug soil pits. May involve some lab work. OS, geology maps etc. might be used.
3 – More complex, in depth and detailed [Data collected over a whole day or longer]	Confident 'soils person'. Familiar with soil process, formation and terminology. Possibly had previous field experience.	Medium to large scale. Soil work and analysis forms the core elements of the proposed study. Possible use of a number of sites, e.g. down a slope or comparison of different land uses.	A range of samples (10–20) and/or soil pits. Range of variables tested using detailed field analysis from available horizons. Comprehensive lab work. OS, land use, soil and geology maps might be used.

Note, the number of samples suggested is merely intended as a guide. The actual number obtained will be determined by the project aims and the precise sampling strategy used.

Soil assessment in the field

Field observations are generally made by either digging a soil pit or by extracting samples of soils using a soil auger (Figure 2.79). Soil pits are most preferable in that they show progressive changes in soil colour with depth (although this is not always the case), but are cumbersome to dig. If you do dig a pit, remove the surface vegetation first and place the dug soil onto a plastic sheet. Ensure that you return the soil to the hole (and the vegetation) once you are finished. Augers are quicker and more convenient to use, but the resulting interpretation of boundaries or horizons might be more difficult to reconstruct. Note

Survey/analysis technique	In the field	Laboratory based
Soil infiltration and degree of compaction	✓	✗
Soil colour	✓	✓
Soil depth (including individual horizons)	✓	✗
Soil erosion risk	✓	✓
Soil moisture	✓	✓
Soil nutrient status	✗	✓
Soil organic content	✗	✓
Soil pH	✓	✓
Soil temperature	✓	✗
Soil texture – finger assessment	✓	✓
Soil texture – sieving, sedimentation and bulk density	✗	✓

Soil survey techniques — Table 2.16

Key: A single tick (✔) is a possible alternative, a cross (✗) indicates not suitable.

Figure 2.79 A soil auger

Soil recording sheet — Table 2.17

SITE _____ SLOPE ANGLE ____° ASPECT _____

SOIL TYPE _____ ROCK TYPE _____

SOIL PROFILE DETAILS

HORIZON	DEPTH (cm)	COLOUR	TEXTURE	pH

that any storage of samples is best done in labelled polythene bags, secured to minimise any moisture loss. If the soil is to be stored for more than a few hours it is best to freeze the sample to minimise any moisture loss.

Table 2.17 is an example of a detailed field soil survey or booking sheet.

Examination of soil horizons (including soil depth)

The various processes which operate in soils (principally controlled by the downward passage of water) may lead to the development of distinct layers or horizons. Sometimes these horizons are exposed, e.g. roadside cuttings or river cliffs, but more often than not, horizons can only be identified by digging a pit (up to 1m in depth should be sufficient). Using an auger will really only work if the soil is damp (so

sticks to the bore) and can be extracted slowly and with some degree of patience. Augers, however, are useful tools in the determination of soil depth.

Different horizons should be recognisable either from differences in colour or texture, or sometimes both. Generally, the deeper the horizon, the stonier it becomes, whilst horizons close to the top of soils are richer in organic matter. Horizons may also display variations in pH and chemistry, e.g. concentrations of sesquioxides (iron and aluminum ions) with depth.

It is good practice to complete a site description for each soil site where samples are taken. Record details such as weather conditions, time of day, type and height of vegetation, land use, drainage conditions, length of slope etc. You should also use large scale OS maps to confirm grid references, site aspect, height, and if possible, geology.

The fieldwork here is straightforward: for each horizon, measure the thickness with a ruler and record its colour. This can be done by either smudging the damp soil onto a piece of white paper, or recording soil colour using a home made coded colour test chart. Compare the soil colour with the printed colour swatch (Figure 2.80) and note the best match.

Soil pH

Soil pH controls the efficiency of mineral and chemical uptake in soil by plants – so it is an important factor in soil suitability for crops. Even if the soil has a good chemical balance, but the pH is unsuitable, the chemicals may not be available for plant uptake. One method for determining pH is to use a chemical 'colorimetric' technique. When the soil sample is mixed with water and an indicator added, a colour change is observed. This colour transformation is then compared with the closest match from a set of predetermined stock samples printed on a test card.

Electronic probes are also available for the analysis of soil pH. Although these may be more convenient to use, bear in mind that they require regular calibration and a fully charged set of batteries to ensure accurate testing. They also tend to be fragile and costly to repair or replace.

Soil texture

This refers to the size of the particles that make up the soil. Three categories are normally used: sand (2 – 0.06 mm), silt (0.06 – 0.004 mm) and clay (<0.004 mm); a 'loamy soil' is a mixture of all three and is preferred for most agricultural purposes.

Figure 2.81 provides a basic method for the examination of soil textures in the field, but as a rule of thumb, sand feels gritty, while silt is smoother and clay tends to be smoother still and sticky.

Soil infiltration and compaction

Infiltration is the passage of water into soil; it is controlled by gravity and capillary action. Different soils (and different soil surfaces) have different infiltration rates. Generally soils which have large particles will be more freely draining than soils which are made primarily of silt and clay. However, soils which have been subject to compaction either by people or machinery, would be expected to have reduced rates of infiltration.

Figure 2.82 shows a diagram of an infiltration tube. This is made from a strong 30 cm length of plastic drainpipe with a standard 10 cm diameter. Remove any surface vegetation and hammer the tube into the ground to ensure a good seal (use a piece of wood over the top to prevent the plastic shattering). Pour water into the tube and measure the rate at which it infiltrates in millimeters per minute. It is important to maintain a regular 'head' of water above the soil, so constant top-ups are required.

Soil compaction can also be determined by using a soil penetrometer. A penetrometer can be home made from a steel pin (knitting needles are ideal or meat skewers) with a bung securely attached to the top. Simply drop the pin from a known height (usually about 1.5 m) into the ground and measure

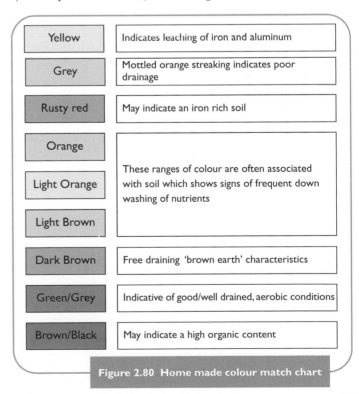

Yellow	Indicates leaching of iron and aluminum
Grey	Mottled orange streaking indicates poor drainage
Rusty red	May indicate an iron rich soil
Orange	These ranges of colour are often associated with soil which shows signs of frequent down washing of nutrients
Light Orange	
Light Brown	
Dark Brown	Free draining 'brown earth' characteristics
Green/Grey	Indicative of good/well drained, aerobic conditions
Brown/Black	May indicate a high organic content

Figure 2.80 Home made colour match chart

Figure 2.81 Field assessment of soil texture

Take a golf-ball sized sample of soil from the appropriate horizon. Moisten with water so it feels damp to the touch and work between thumb and forefinger. Working down the list of statements below, record the texture according to the most appropriate description.

1. Does the soil form a coherent ball?
Easily . . . . (move to test 2)
With great care . . . . LOAMY SAND
But check using tests 1 and 2
No . . . . SAND

2. What happens when the ball is pressed between thumb and forefinger?
Flattens coherently . . . . (test 3)
Tends to break up . . . . SANDY LOAM
But check using tests 3 and 4

3. On slight further moistening can the ball be rolled into a cylinder about 5 mm thick?
Yes . . . . (test 4)
No, the ball collapses . . . . SANDY LOAM

4. On slight further moistening can the ball be rolled into a thinner cylinder about 2 mm thick?
Yes . . . . (test 5)
No . . . . SANDY LOAM

5. Can the thread be bent into a horseshoe around the back of the hand?
Yes . . . . (test 7)
No . . . . (test 6)

6. On remoulding with further moisture what does the soil now feel like?
Smooth and pasty . . . . SILT LOAM
Rough and abrasive . . . . SANDY SILT LOAM

7. Without cracking, can a ring of about 25 mm in diameter be formed by joining the ends of the thread?
Yes . . . . (test 9)
No . . . . (test 8)

8. On remoulding with additional water, what does the soil now feel like?
Very gritty . . . . SANDY CLAY LOAM
Moderately tough . . . . CLAY LOAM
Doughy . . . . SILTY CLAY LOAM

9. On remoulding again (without re-wetting), can the surface be polished with a finger?
Yes, with a high polish . . . . (test 10)
Yes, but there are noticeably gritty particles . . . . SANDY CLAY
No . . . . (go back and retry test 8)

10. On wetting thoroughly does the soil stick fingers together?
Yes, very strongly . . . . CLAY
Yes, but only slightly . . . . SILTY CLAY

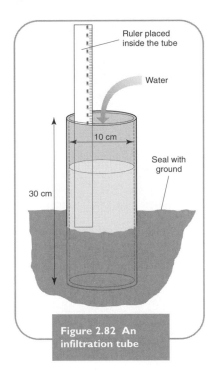

Ruler placed inside the tube

Water

10 cm

Seal with ground

30 cm

Figure 2.82 An infiltration tube

the depth to which the pin penetrates. This technique is often used to record compaction across a footpath where the results can be displayed as an inverted histogram.

Soil temperature

Soil temperature controls seed germination and speed of plant growth. It may also give an indication of soil texture – sandy soils tend to heat up more easily and lose heat more quickly than heavier, clay soils. Measure soil temperature at set depths using a protected thermometer or special probe. Expect soil temperatures to fluctuate less with increasing depth. This type of technique is best suited to investigations where temperature is measured over time, for example, every hour over a 24 hour period.

Laboratory-based soil analysis

Remember that there are a range of additional soil analyses that can be carried out back at school in a laboratory. These include the following.

▶ Soil moisture content

▶ Soil organic content

▶ Soil nutrient status

▶ Soil texture: sieving, sedimentation and bulk density.

Water content

To calculate the water content of a soil sample, place a small, weighed sample of soil (10–20 grams is sufficient) into a heat resistant ceramic pot and oven dry for a minimum of 2 hrs at 100°C. Reweigh the sample and calculate the percentage moisture loss.

$$\% \text{ moisture} = \frac{(\text{raw weight} - \text{dry weight})}{\text{dry weight}} \times 100$$

Organic content

To estimate soil organic content, use the same oven dried soil and a suitable container, e.g. a crucible and burn off the organic material. This requires heating the soil to 600°C for about 20–30 minutes (using a bunsen burner). Then calculate the percentage organic content:

$$\% \text{ organic content} = \frac{(\text{dry weight} - \text{burnt weight})}{\text{dry weight}} \times 100$$

Safety-tip

You must wear safety goggles and heat resistant gloves.

Nutrient testing

Nutrient testing kits contain fluids which test soils for nitrates, phosphates and potassium in addition to lime. Soil nitrates can also be determined using nitrate test strips, although the nitrate must first be leached from the soil: mix 50 g soil with 75 ml of water and filter the mixture. Nitrate concentrations will vary seasonally and spatially. Rainwater, for example, can flush nitrates downhill, so nitrate concentrations are often high at the base of hills.

Soil sieves

These provide a more quantitative technique to determine soil texture rather than the somewhat subjective field texture test. Although expensive to buy, they accurately divide the soil into different particle or phi sizes. The weight and volume of each particle size can then be calculated. Plot the data as a divided histogram.

Sedimentation

This is a simple approach that relies on the fact that larger particles have a faster settling velocity compared to smaller and finer particles. Add approximately 100 g of separated soil to a tall measuring cylinder and fill it with water. Shake the container (with your hand over the end) and leave to settle for up to three days. Using the graduations on the side of the cylinder read off the depths of the different bands (Figure 2.83) and express each class as a percentage of the depth of the soil in the cylinder.

Figure 2.84 shows a soil texture (triangular) graph, whereby a description of soil texture can be given based on the relative proportions of sand, silt and clay.

Bulk density

The bulk density of a soil is a function of the proportions of mineral and organic matter, soil compaction and soil texture. This measure is strongly affected by land use and soil management. Bulk density can be obtained by extracting soil in an infiltration tube. Oven dry the soil overnight at 110°C and then calculate the bulk density.

Generally the higher the bulk density result, the more compacted the soil. The average bulk density of a non compacted, loamy soil is around 1.25 g/cm³, whilst peat has a density of about 0.5 g/cm³.

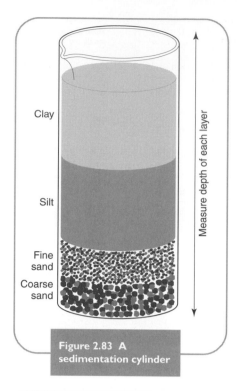

Figure 2.83 A sedimentation cylinder

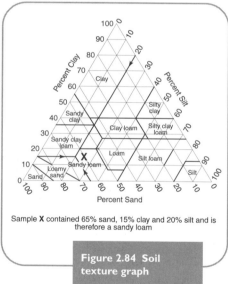

Sample **X** contained 65% sand, 15% clay and 20% silt and is therefore a sandy loam

Figure 2.84 Soil texture graph

TOP TIPS

There are soil records for most parts of Britain which may provide relevant data to enhance a soil study. It may be possible, for example, to make comparisons between your data and published Soil Survey maps. You could comment on the reliability of such maps and their accuracy.

$$\text{Bulk density} = \frac{\text{weight of dry soil (g)}}{\text{volume of soil in tube (cc)}}$$

For example, for compacted soil,
weight = 800g
Tube Radius = 5 cm
'Depth' of soil = 30 cm
So volume of soil = pi*r*2d
Pi = 3.14, r = 5, d = 30
Volume = 471 cm³
800/471 = 1.70 g/cm³ Bulk density

Soil erosion risk

Various factors control soil erosion risk and susceptibility:

▶ slope

▶ plant cover

▶ soil texture

▶ rainfall intensity

▶ wind speed

▶ degree of exposure.

A score can be developed based on these factors to establish a soil erosion risk factor. Try and determine a soil risk factor for contrasting areas under different weather events, e.g. light rain v downpours, or windy and dry conditions.

Table 2.18 provides a simple classification of soil erosion for *bare soil surfaces*.

Preparation of a soil risk map will identify the most vulnerable areas of a site (Figure 2.85). The map should include valley features that can channel water, and identify where erosion risk may be increased locally. Gateways, roads and ditches should be marked as these can be sites of deposition or erosion.

	Soil erosion risk for bare soil surfaces			Table 2.18

Soil texture	Steep slopes (>7°)	Moderate slopes (3–7°)	Gentle slopes (2–3°)	Flat ground (<2°)
Sand, loamy sand, sand loam, sand silt loam, silt loam	Very High	High	Lower	Slight
Silty clay loam	High	Moderate	Lower	Slight
Other mineral soils	Lower	Slight	Slight	Slight

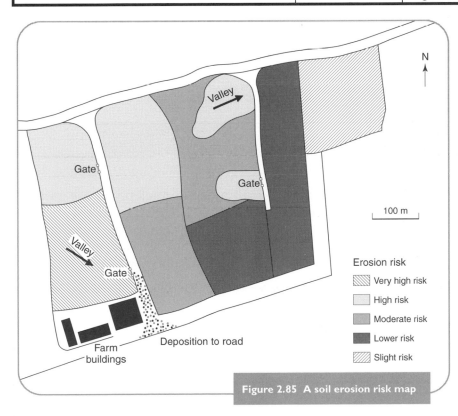

Erosion risk

▨ Very high risk

☐ High risk

▓ Moderate risk

�－ Lower risk

▨ Slight risk

Figure 2.85 A soil erosion risk map

TOP TIPS

Why not suggest measures to reduce soil erosion? For agricultural land these could include:

1. Alterations to the farm layout; relocations of field entrances to avoid deposition of sediment on roads or into ditches and water courses.

2. Adjustments to rotations and cropping/land uses. For example, in higher risk fields switch from late sown autumn to spring crops.

3. Adoption of good management practices for erosion control – maintain crop cover, increase soil organic matter and avoid overworking/over-cropping the land, plant hedges.

In additional to the above, consider preparing your own soil erosion management plan.

INVESTIGATING VEGETATION

Description of vegetation	Quadrat size: length of side(m)
Lichens and mosses	0.25×0.25
Grasslands and small vegetation	0.5×0.5
Tall herbs and heaths	1–2
Scrub and woodland shrubs	5–10
Woodland canopies	25–50

Vegetation survey techniques

A quadrat is used to define a small area of vegetation for ecological study. It is usually a convenient square frame used to determine the abundance of plants within this area. Plants are recorded by their percentage cover, frequency or density. There are three main types of quadrat:

▶ **Frame quadrats** (Figure 2.86) These can be home made using four pegs inserted into the ground and wrapping string around them. Alternatively frames can be constructed from approximately 15 mm diameter doweling or plastic pipe.

▶ **Grid quadrats** (Figure 2.86) These are simply frame quadrats subdivided into smaller units, either 25 or 100.

Figure 2.86 A grid quadrat and frame quadrat

The size of plants or vegetation being studied should control the size of quadrat used.

▶ Where the vegetation is tall or multi-layered (as in a grassland), **point quadrats** may be more suitable as they are less likely to flatten the vegetation underneath them (Figure 2.87).

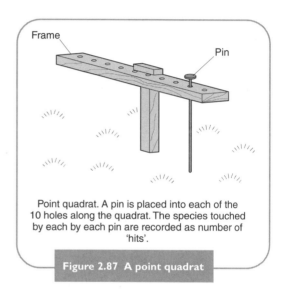

Point quadrat. A pin is placed into each of the 10 holes along the quadrat. The species touched by each by each pin are recorded as number of 'hits'.

Figure 2.87 A point quadrat

Ten pins are dropped through holes in a narrow frame and if the pin touches a plant it is recorded as a 'hit'. The data for the number of hits per quadrat can then be converted to percentage frequency data (see below).

$$\% \text{ frequency of species A} = \frac{\text{number of hits for A}}{\text{maximum number of possible hits}} \times 100$$

If there are 10 holes in the frame, the maximum number of hits will be 10, ie 100%

TOP TIPS

Choose your quadrat type carefully as they each have a number of advantages and disadvantages:

Quadrat Type	Advantages	Disadvantages
Frame	Light to carry, do not tend to squash the vegetation	Difficult to estimate and cover quickly
Grid	Commonly held at schools, easy to use	Not ideal for tall vegetation
Point	Enables full vegetation structure to be observed	Specialist equipment may not be readily available

Recording vegetation types

There are a number of different strategies available for recording vegetation as follows.

▶ **Percentage cover** When using a frame or grid quadrat, percentage cover of species is usually recorded. This involves estimation of the ground area covered by each species present (which is often easier with grid quadrats). Layering of vegetation may result in a total cover which exceeds 100 per cent.

▶ **Frequency** The number of quadrats in which the species are found is expressed as a percentage of the total number of samples.

$$\% \text{ frequency of species A} = \frac{\text{number of quadrats containing species A}}{\text{total number of quadrats surveyed}} \times 100$$

▶ **Abundance** Use different categories to classify the number (or percentage cover) of a species. The ACFOR scale assigns a letter to each category of abundance (Table 2.19).

The ACFOR scale | Table 2.19

Letter	Description	Amount of cover (approx %)
A	Abundant	>50
C	Common	25 – 50
F	Frequent	10 – 25
O	Occasional	1 – 10
R	Rare	<1

▶ **Diversity** The total number of different types of species present in a sampling area or quadrat can be used as a gauge of the vegetation diversity.

Simpsons Diversity Index can be used as shown on the top of the next page.

▶ **Height** This may be a useful indicator when determining the impact of recreational pressure on a particular resource. Some plants, e.g. heather, can be damaged by trampling and show a reduction in height as a result.

TOP TIPS

For non specialists, plant identification can be difficult, especially in the winter months when plants are not in flower. For 'difficult' species, collect leaf samples for identification back at home or school.

Diversity $(D) = \dfrac{N(N-1)}{\sum n(n-1)}$

Where D = diversity index, N = total number of individual plants, n = number of individuals per species. $\sum$ = the 'sum'

Example:

Plants recorded in a 2 × 2m quadrat area.

An advantage of this test is that you don't need to know all the plant types, just recognise differences!

Species	Number
Daisy	15
Ragwort	42
Nettle	10
A – not identified	102
B – not identified	2
C – not identified	61
Total $\sum$	232

$$D = \frac{232\,(232-1)}{15\,(15-1) + 42\,(42-1) + 10\,(10-1)\ \text{etc...}} = \frac{53{,}592}{15{,}986}$$

D = 3.35

A lower 'D' value indicates a lower plant diversity and perhaps a more unstable environment or community. More mature ecosystems show a greater diversity and a higher D value.

How many samples?

The number of samples required for a vegetation survey will depend on site diversity, site sensitivity and the time available. As a rule of thumb, the more species present, the more samples required. However, it is possible to make a more scientific estimation of the number of samples needed. First count the number of individuals of the species you are studying within a random sample of 20 quadrats. Then calculate the running or cumulative mean of the data:

Sample number	Number of species X	Running mean
1	5	(5/1) = 5.00
2	8	((8+5)/2) = 6.50
3	4	((4+8+5)/3) = 5.67
4	2	((2+4+8+5)/4) = 4.75
5	2	((2+2+4+8+5)/5) = 4.20
etc (up to 20)		

The running mean can then be plotted on a graph (Figure 2.88). Where the graph starts to flatten to a fairly stable and non-fluctuating mean, this is the minimum number of samples for species X at that site.

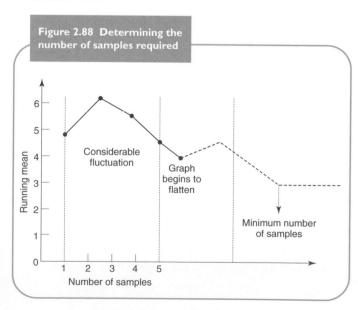

Figure 2.88 Determining the number of samples required

TOP TIPS

Choosing the right project

Ecosystem studies normally follow one of four main routes to enquiry:

1. Vegetation species type, diversity and dominancy.

2. Percentage cover of vegetation, including bare ground, height, density, layering and structure.

3. How the vegetation has changed over time (succession), how the species have adapted to the physical environment and might only be found in certain locations (zonation), and to what extent human activity has modified the habitat.

4. An evaluation of the success/failure of current or past management practices.

More example investigations are given at the end of each ecosystem type.

COMMON BRITISH ECOSYSTEMS

Sand dune ecosystems

Sand dunes are accumulations of sand grains, shaped into mounds or ridges by the wind under the influence of gravity. Macro-tidal environments combined with onshore winds and wind-driven currents provide the ideal conditions for dune development. Coastal sand dunes are diverse ecosystems, involving a complex interaction between biotic plant communities and environmental or abiotic conditions. Dunes are also dynamic systems providing evidence for phases of building and erosion, varying nutrient status, past and present management, local climate and topography.

Sand dune studies usually follow one of two major themes (although many investigations combine elements of both aspects).

▶ Changes in the vegetation, i.e. structure, type and composition, across the sand dune from the water's edge to inland. This type of study is normally used to investigate (or replicate) succession.

▶ An assessment of the people-environment and/or human impact on the sand dune resource. This also can include sand dune management.

It is usual to use a transect to study sand dune zonation and succession. Starting from the strand line (nearest the sea), a tape is set down and appropriate sampling intervals are determined. At each point a number of biotic variables can be measured: percentage cover of plants, plant height, soil type, depth and pH etc. During the design stage, it is also important to consider the following.

▶ Whether the distance between sampling points is regular (a systematic approach), or stratified, whereby the interval distances are modified according to the (abiotic/biotic) changes taking place. If there are few changes in the data collected the interval can be extended,

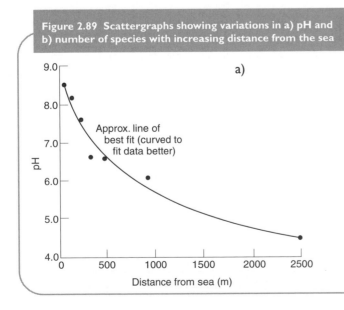

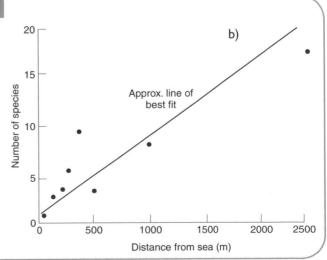

Figure 2.89 Scattergraphs showing variations in a) pH and b) number of species with increasing distance from the sea

but if many changes are observed in a short space of time then the sampling interval should be shortened to take account of such changes.

▶ Which other biotic variables are going to be measured, e.g. gradient, infiltration rates or microclimate.

In sand dune environments it may be possible to establish relationships between variables (pH, soil depth, number of species and plant ages determined from old maps) and distance from the sea (Figure 2.89). Sand dune data can also be displayed as a composite diagram (Figure 2.90).

Hydroseres

The term hydrosere refers to plant succession in a fresh water environment, e.g. around a pond or lakeshore. A study which considers changes in the vegetation

structure, type and composition can be undertaken in such an environment. Either a stratified or systematic sampling approach will be appropriate, from the water's edge to 10–20 m inland (or where there is no change in vegetation type as a result of increasing distance from the water's edge).

TOP TIPS

Start with an 'ecological' sketch map of the lake/pond, including a full key which identifies the main environments and habitat types (Figure 2.90).

Along your transect, measure the depth of water and the height of plants, as well as taking soil samples from a constant depth in order to establish any changes in organic content. This data should be

---

Figure 2.90 Sand dune data

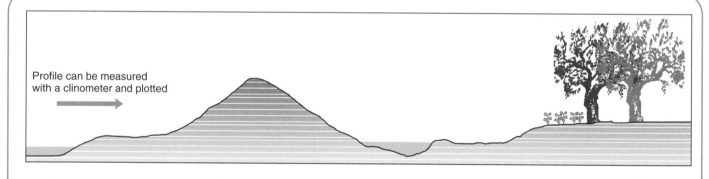

Profile can be measured with a clinometer and plotted

Site number/description	Strand Line	Embryo Dunes	Fore Dunes	White or Yellow Dunes	Fixed Dunes	Dune Slack	Dune Scrub	Dune Heath	Coniferous Woodland
Distance from the sea (m)	0–20	20–80	80–150	150–300	300–500	500–600	650–700	700–2500	2500+
Approximate age (years)	0								100+
Main soil colour	–	Yellow			Yellow/Grey		Grey	Brown	Grey
Soil surface pH	8.5	8.0	7.5	7.0	6.5		6.0		5
% Calcium carbonate	10	8	8	5	1		<0.1		
% Humus	<1			2.5	5	10	12		>40
Mean soil depth (cm)	0	0	0	0	3	42	40	41	23
Mean slope angle (°)	1	5	15	8	11	6	3	2	2
Mean % cover marram grass	0	15	30	70	55	0	0	0	0
Mean % cover heather	0	0	0	0	0	0	25	80	15
No. of species	1	3	4	7	9	12	4	8	6

plotted in an appropriate way, (Figure 2.91). Scatter graphs can be produced to illustrate how different variables change with distance along the transect, e.g. soil pH or moisture (similar to Figure 2.89) Provided there is enough data (a minimum of 10 pairs), test your results for statistical significance using the Spearmans' Rank test.

In addition you could explore the degree of thermal stratification in the water body and try to relate this to lake/pond flora and fauna (i.e. invertebrates).

Figure 2.92 shows a depth temperature graph of a lake in Shropshire. This data can be collected with a waterproof thermometer connected to sticks or poles (i.e. bamboo canes) of various lengths. Ideally a thermometer which is well insulated (similar to ones used for the measurement of soil and ground temperature) should be used to minimise fluctuations in temperature. Collect macro-invertebrates at different depths with a sampling net as described on page 45.

Hydroseres are often well used and valuable resources. An assessment of the people-environment and human impact on the site could prove an interesting option, i.e. activity surveys (see page 206) or footpath transects (see page 127). You may also consider a study which evaluates the limitations/successes of management strategies.

Salt marshes

Salt marshes offer a good alternative to sand dunes for the study of succession and people-environment themes. Salt marshes

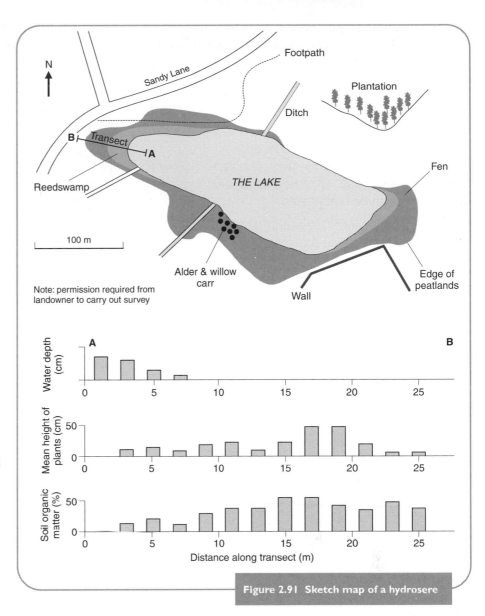

Figure 2.91 Sketch map of a hydrosere

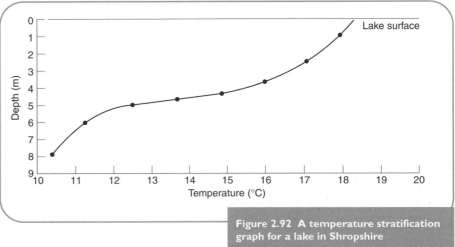

Figure 2.92 A temperature stratification graph for a lake in Shropshire

develop on sheltered muddy shores – the build up of muds through coagulation and sedimentation raises the height of the salt marsh relative to the sea so that daily immersion no longer occurs. The subsequent vegetation structure (Figure 2.93) forms as a result of the environmental gradient created by the more infrequent washing of the tidal sea across the salt marsh. There are various studies that can be considered.

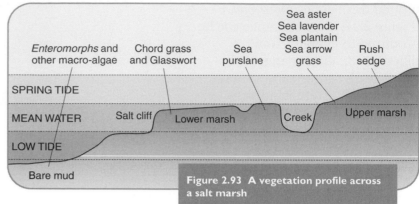

Figure 2.93 A vegetation profile across a salt marsh

▶ Vegetation surveys. Use a stratified, interrupted belt transect at suitable height intervals up the salt marsh (probably 25–50 cm). It is preferable to survey at vertical height intervals rather than horizontal distance intervals, since the vertical movement of the tides is the main controlling factor on vegetation structure in this environment. Use a 0.5 × 0.5 m quadrat as described on page 112. Figure 2.94 shows kite diagrams for an interrupted transect in Pembrokeshire.

▶ The soils can be observed at each sampling point and profile diagrams constructed. Soil examination can be based on colour, depth of root zone, depth of anaerobic (black) muds and horizons. The removal of a small soil sample in the field enables changes in

pH, organic matter and salinity levels to be determined back in the laboratory at school.

Other ideas for further study include the following.

▶ Seasonal changes in salt marsh vegetation.

▶ Plant adaptations to limiting factors.

▶ Salinity variations in salt marsh soils.

▶ Vegetation zonation with distance from a salt pan or drainage gully.

▶ Impact of trampling on salt marsh vegetation.

▶ Effects of animal grazing on salt marsh vegetation.

▶ Reclamation and conflict land uses on salt marshes for agriculture and building.

Upland ecosystems

In Britain moorland ecosystems can occur above 250 metres in altitude, mainly on resistant impervious

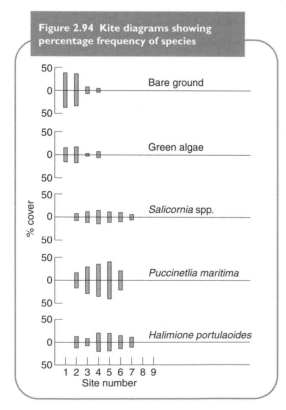

Figure 2.94 Kite diagrams showing percentage frequency of species

These types of ecosystem are of low primary productivity and low diversity dominated by heather. As the rocks in upland areas are mostly acidic they therefore supply few nutrients to the soil, which combines with the climatic conditions to provide an environment which restricts plant growth.

rocks of upland Britain, such as the Millstone grit of the Pennines.

Upland areas can provide a range of studies often based on the relationship of ecosystem structure to management regimes. Table 2.19 is an example recording sheet for use in a ecosystem survey of an upland area. Comparative vegetation studies can be made by looking at different age stands of heather which result from management intervention (either burning or cutting). The use of

A recording sheet for an upland ecosystem — Table 2.19

	Species	Age of stand (yrs)				
		1	6	9	13	20
Mean % Cover*	Heather	45	60	90	100	95
	Bilberry	75	60	30	25	30
	Cowberry	45	50	50	30	25
	Wavy hair grass	90	40	60	50	80
	Bracken	20	0	0	0	65
	Gorse	30	0	0	0	70
[blank spaces for additonal species]						
	Moses and lichens	30	25	20	10	15
	Bare earth and leaf litter	40	0	0	0	0
Mean Height (cm)	Heather	20	31	58	74	70
	Bilberry	21	18	11	12	22
Biomass Heather (gm collected)		218	314	284	409	357
Biomass Green Shoots (gm)		159	109	73	51	22
% Green Shoots		72	35	26	12	6
Light at ground level as a % of that on the top of heather canopy		92	49	31	22	18

*Note mean % cover can add up to more than 100% because of the different layers of vegetation.

a simple heather life cycle model (Figure 2.95), for instance, allows a comparative analysis of different ecosystem aspects.

Another aspect to consider might be the ratio of total heather biomass, compared to the biomass of heather green shoots at different ages through the life cycle. To do this, cuttings of heather would need to be taken, weighed and then the green shoots removed and weighed separately. Sketch maps can be drawn of the site displaying ages of stands, mean height of plants and percentage of green heather shoot.

$$\text{% Green shoots} = \frac{\text{Mass of green shoots (g)}}{\text{Total plant biomass (g)}} \times 100$$

More ecologically focused research can be centred around plant association investigations. Many plants often grow very close to only a few other species, heather and bilberry for instance. A simple method of identifying such associations is to set out a large taped quadrat measuring approximately 40×40 m. Using a small 25×25 cm quadrat, randomly select 100 locations within the larger frame and record (using a tally) the presence of any species within the small quadrat. Go through the results and every time two plants are found in the same quadrat join their names with a line of 1 mm thickness. Those plants in close association will stand out as being joined together by thicker black lines. In the example (Figure 2.96), there is a close association between bilberry and heather.

Figure 2.95 Heather life cycle model

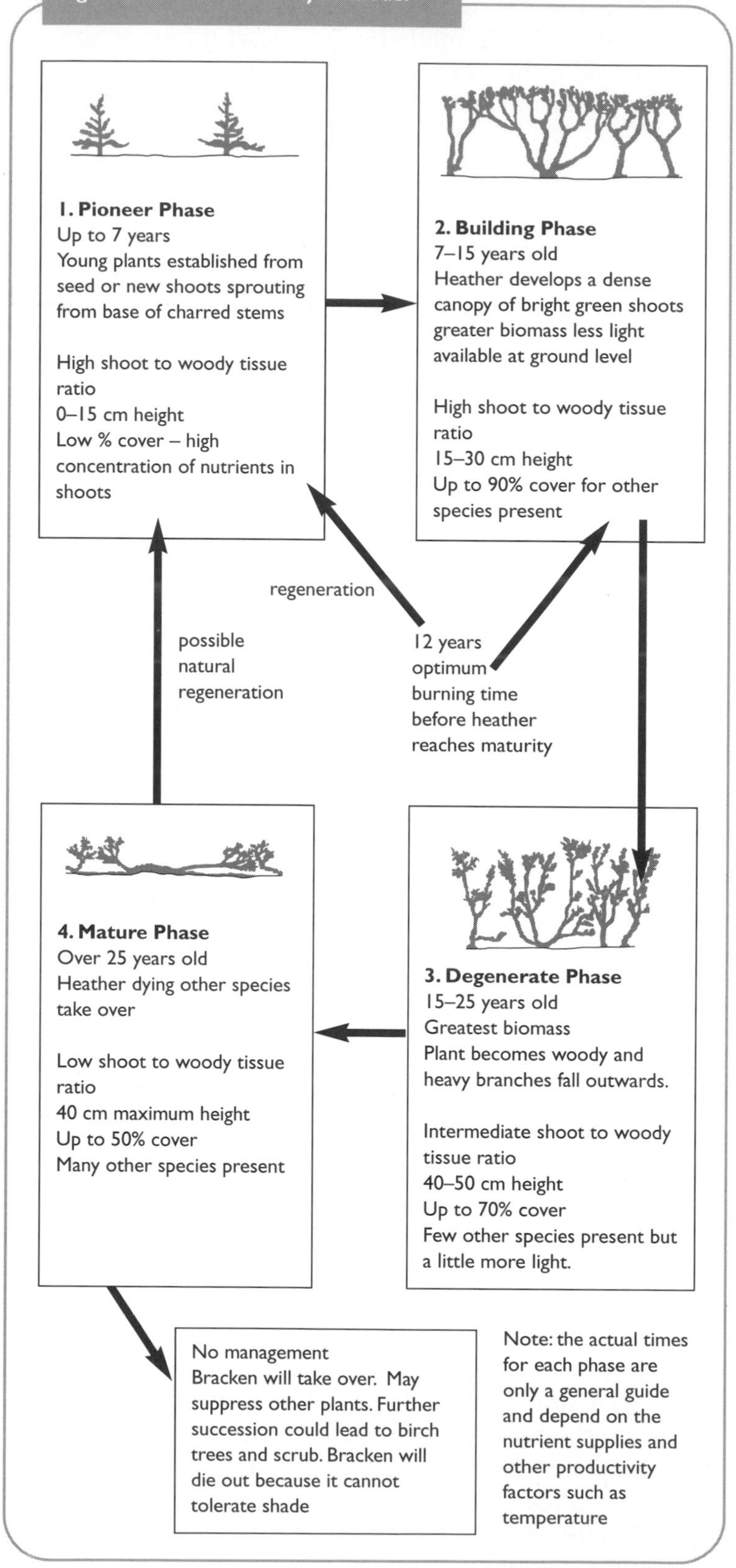

1. Pioneer Phase
Up to 7 years
Young plants established from
seed or new shoots sprouting
from base of charred stems

High shoot to woody tissue
ratio
0–15 cm height
Low % cover – high
concentration of nutrients in
shoots

2. Building Phase
7–15 years old
Heather develops a dense
canopy of bright green shoots
greater biomass less light
available at ground level

High shoot to woody tissue
ratio
15–30 cm height
Up to 90% cover for other
species present

regeneration

possible
natural
regeneration

12 years
optimum
burning time
before heather
reaches maturity

4. Mature Phase
Over 25 years old
Heather dying other species
take over

Low shoot to woody tissue
ratio
40 cm maximum height
Up to 50% cover
Many other species present

3. Degenerate Phase
15–25 years old
Greatest biomass
Plant becomes woody and
heavy branches fall outwards.

Intermediate shoot to woody
tissue ratio
40–50 cm height
Up to 70% cover
Few other species present but
a little more light.

No management
Bracken will take over. May
suppress other plants. Further
succession could lead to birch
trees and scrub. Bracken will
die out because it cannot
tolerate shade

Note: the actual times
for each phase are
only a general guide
and depend on the
nutrient supplies and
other productivity
factors such as
temperature

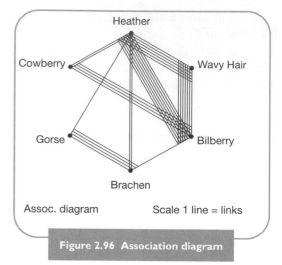

Heather

Cowberry   Wavy Hair

Gorse   Bilberry

Brachen

Assoc. diagram   Scale 1 line = links

Figure 2.96 Association diagram

TOP TIPS

Vegetation association can also be
described statistically using the chi-
squared test. The null hypotheses might be:
There is no association between Species A
and Species B.

It is also possible to carry out an association
index test:

$$I = \frac{\text{No. of species in area A}}{\text{No. of species in area B}} \times 100$$

**A high percentage index indicates that the
areas have a lot of similarities in terms of
number of species.**

Other upland ecosystem studies may
concentrate on the impact of people on the
ecosystem or an evaluation of the degree of
success relating to management of the site.

Woodlands

Woodlands are a vital component of the landscape and have, throughout history, been important to people. During the Middle Ages, for instance, British woodlands provided timber, firewood, many types of food including wild animals and birds (game), nuts, berries and fungi, as well as grazing land for domesticated farm animals. Woodlands today are used for commercial timber production and outdoor recreation activities. In Britain there are three main woodland types:

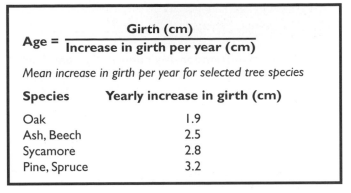

$$Age = \frac{Girth~(cm)}{Increase~in~girth~per~year~(cm)}$$

Mean increase in girth per year for selected tree species

Species	Yearly increase in girth (cm)
Oak	1.9
Ash, Beech	2.5
Sycamore	2.8
Pine, Spruce	3.2

▶ **Ancient woodland** which has a historical record dating back to pre 1600AD. Indicator plants for this type of woodland include: small leaved lime, sessile oak, wood anemone and dogs mercury.

▶ **Semi-natural woodland** – younger than ancient woodland, this ecosystem has regenerated from seeds dispersed by plants. A typical indicator plant for this type of wood is birch.

▶ **Plantation woodland** can either comprise broad-leaved or coniferous trees. The latter are often found on marginal uplands where the environmental conditions are too harsh for native broad-leaves to do well.

Woodland areas are accessible to everyone – even large cities have suitable study sites. Research into woodland ecology can provide data that support spatial analysis and also an opportunity to examine people-environment activity. Most woodland surveys require a sketch or base map of the area and comparisons to be made between survey sites in the same woodland or two nearby woodlands, e.g. deciduous v coniferous. Often sampling is based on a 30 × 30 m sampling frame. Project ideas are outlined below.

▶ Identify and record the general distribution of the main tree species and stands – look specifically at dominancy and density.

▶ For selected tree species measure their girth (at waist height) and height and estimate their age.

▶ Measure the distance from each of the selected trees to its nearest neighbour (both for the same species and different types). Can this be related to height and girth?

▶ Record light intensity and plot your results as an isopleth map (Figure 2.97). Is there a relationship between this and the species type and age of trees? Light intensity can also be approximated using a quadrat held above the head.

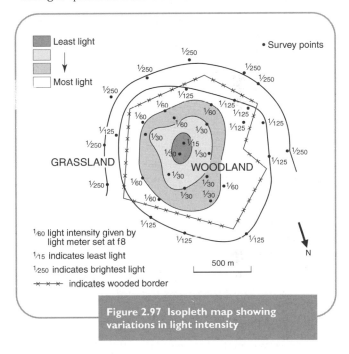

Figure 2.97 Isopleth map showing variations in light intensity

▶ Record types of visitor and reasons for their visit: visitor profiles and attitudes can be examined through the use of a suitably designed questionnaire. Figure 2.98 provides an example.

▶ Assess the impact of recreation/human pressure on a site. This can be linked to ecological impact (Table 2.20) and the degree of management (Table 2.21) with declining distance from entrances and exits to the woodland.

Figure 2.98 Woodland visitor questionnaire

Good morning/afternoon. My name is ...... and I am from .... Can I ask you a few questions about your reasons for your visit today? It should only take 2–3 minutes.

1. Where have you come from today? (include postcode)

   _____
   _____
   _____

2. How did you get here? (*tick all that apply*)

   Car ☐   Other ☐

   Bus ☐

   Cycle ☐

   Train ☐

   Foot ☐

3. Have you been here before?

   No ☐

   Once ☐

   Few times ☐

   Many times ☐

4. What is the reason for your visit? (*tick all that apply*)

   Walking ☐   Bird watching ☐

   Horse riding ☐   Cycling ☐

   Picnicking ☐   Other ☐

   Photography ☐

5. Do you know who owns the site?

   Yes ☐   No ☐

6. Have you looked at any of the information boards?

   Yes ☐

   No ☐

   Not sure ☐

7. Do you regard this area as being:

Clean	7 6 5 4 3 2 1	Dirty
Noisy	1 2 3 4 5 6 7	Quiet
Beautiful	7 6 5 4 3 2 1	Ugly
Boring	1 2 3 4 5 6 7	Interesting
Open	7 6 5 4 3 2 1	Enclosed
Ordinary	1 2 3 4 5 6 7	Spectacular
Natural	7 6 5 4 3 2 1	Unspoilt
Monotonous	1 2 3 4 5 6 7	Varied
Good order	7 6 5 4 3 2 1	Neglected

8. What management schemes from the list below have you seen on your visit today? (*tick all that apply*)

   Litter bins ☐

   Gates/styles ☐

   Noticeboards ☐

   Fenced off areas ☐

   Harvesting/coppicing ☐

9. Do you consider the facilities here to be:

   Poor ☐

   Adequate ☐

   Good ☐

   Excellent ☐

10. Would you like to suggest any improvements?

    _____
    _____
    _____
    _____
    _____

Male	Female	Approximate age

The index for ecological impact | Table 2.20

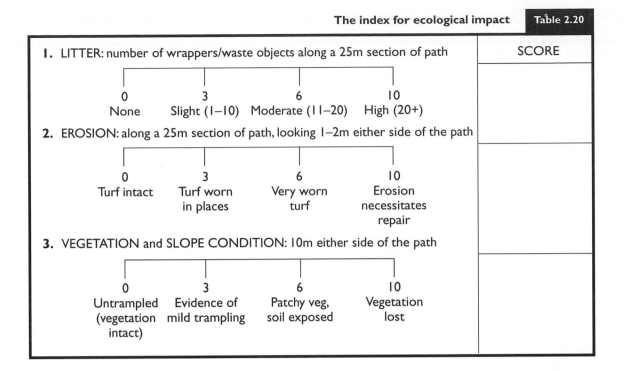

	SCORE
1. LITTER: number of wrappers/waste objects along a 25m section of path	
0 — None / 3 — Slight (1–10) / 6 — Moderate (11–20) / 10 — High (20+)	
2. EROSION: along a 25m section of path, looking 1–2m either side of the path	
0 — Turf intact / 3 — Turf worn in places / 6 — Very worn turf / 10 — Erosion necessitates repair	
3. VEGETATION and SLOPE CONDITION: 10m either side of the path	
0 — Untrampled (vegetation intact) / 3 — Evidence of mild trampling / 6 — Patchy veg, soil exposed / 10 — Vegetation lost	

The index for management | Table 2.21

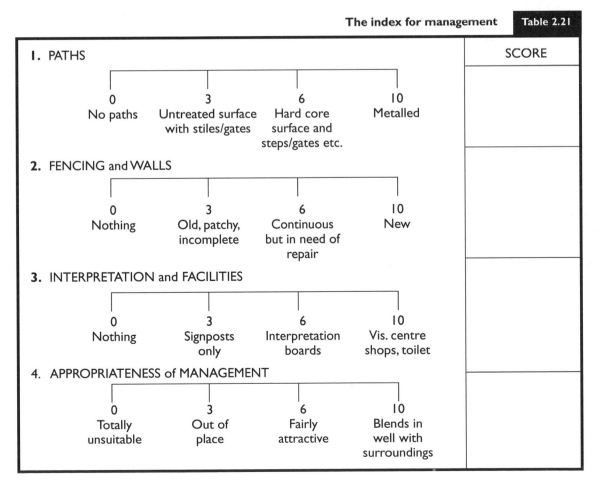

	SCORE
1. PATHS	
0 — No paths / 3 — Untreated surface with stiles/gates / 6 — Hard core surface and steps/gates etc. / 10 — Metalled	
2. FENCING and WALLS	
0 — Nothing / 3 — Old, patchy, incomplete / 6 — Continuous but in need of repair / 10 — New	
3. INTERPRETATION and FACILITIES	
0 — Nothing / 3 — Signposts only / 6 — Interpretation boards / 10 — Vis. centre shops, toilet	
4. APPROPRIATENESS of MANAGEMENT	
0 — Totally unsuitable / 3 — Out of place / 6 — Fairly attractive / 10 — Blends in well with surroundings	

To calculate the indexes for ecological impact and management, use the following formula:

$$\text{Index} = \frac{\text{total score}}{\text{number of variables}}$$

For example, if the total score for a site's ecological impact = 19, and there are 3 variables being measured, the index = 19/3 = 6.3

A high score for ecological impact indicates a reduced quality site (probably in poor condition), whilst a high score for management implies a site which is in better condition, in terms of its management. To use the indexes effectively comparisons must be made between different areas or different sites.

TOP TIPS

Isoline maps can be used to depict woodland ecology, conservation, management and visual quality as well as the biogeography of the wood. Statistical analysis is also possible using Spearman's Rank to examine relationships between such variables as pH and percentage tree cover, or ecological impact and distance from various entrances/exit points to the wood. Where two different age stands are being compared, the chi-squared test can be used to identify statistically valid similarities or differences.

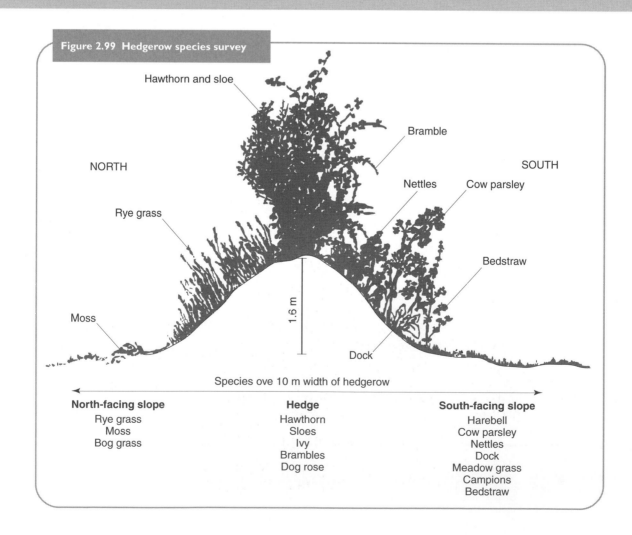

Figure 2.99 Hedgerow species survey

Hawthorn and sloe

Bramble

NORTH   SOUTH

Nettles   Cow parsley

Rye grass

Bedstraw

1.6 m

Moss

Dock

Species ove 10 m width of hedgerow

North-facing slope	Hedge	South-facing slope
Rye grass	Hawthorn	Harebell
Moss	Sloes	Cow parsley
Bog grass	Ivy	Nettles
	Brambles	Dock
	Dog rose	Meadow grass
		Campions
		Bedstraw

Hedgerows

Hedgerow environments offer a range of straightforward studies. Firstly select a suitable length of hedge (20–40 m) and record, using a suitable sampling strategy (perhaps every 5 m) the height, depth/width and condition. Hedge condition or quality should be determined with a subjective scoring system using a scale of 0–5.

At more frequent intervals (1–2 m), record the hedge vegetation. This can either be done using a 0.5 × 0.5 m quadrat or smaller, or with point sampling. Look for changes in plant type and cover/frequency along your length of hedge, and also changes with structure or height within the hedge, or compare one side of the hedge with another. Figure 2.99 shows the hedge species diversity that results from different aspects and conditions.

Another idea is to try and relate the age of a hedge to species diversity. As a rule of thumb, the older the hedge, the greater the diversity. Secondary data in the form of historical maps are required to determine the approximate age of the hedge. Results can be plotted as a scatter diagram (Figure 2.100)

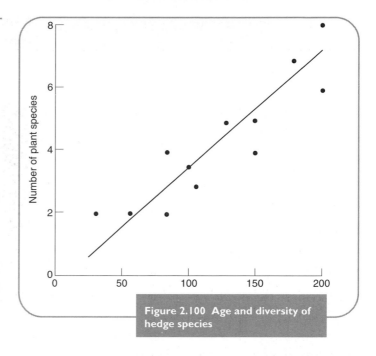

Figure 2.100 Age and diversity of hedge species

TOP TIPS

In many parts of Britain hedges have been removed to create larger fields, or replaced by fences. Use historical maps, together with your own primary data, to investigate these changes through time. When, for example, were the greatest hedgerow losses? Do hedgerow losses show any patterns spatially? What incentives and strategies are now being used to slow or even reverse the trend in hedgerow removal?

Gardens

Gardens provide a range of opportunities for personal investigations. An obvious starting point is to measure the size of the garden (either with a tape or by pacing) and establish its area in m². Produce an annotated map indicating any slopes, boundaries and notable features, e.g. bare soil surfaces. Also survey and record the variety of plants found in the garden – try to determine whether they are natural (i.e. not intentionally planted) or have been deliberately planted to improve the appearance of the garden. Try to evaluate whether the different areas within the garden are being maximised to their full potential. In other words are the correct types of plants being sewn in an appropriate area to maximise aesthetic quality and their tolerance of the soil conditions which dominate a particular region of the garden?

Similar approaches to fieldwork that are used in 'garden surveys' can be transferred to other locations such as the school grounds or a local park/ amenity area. There are a number of alternative topics that could also be researched in these type of environments.

▶ An evaluation of where to put plants to create a more sheltered environment.

▶ Locating the most appropriate areas for seating, recreational use or a planted garden.

ASSESSING THE HUMAN IMPACT ON ECOSYSTEMS

Human impact surveys on ecosystems are normally based on the following assumptions:

▶ Vegetation development, structure and type is affected by recreational pressure.

▶ Recreational pressure can vary according to the time of day, week or season.

▶ Pressure of people is focused in and around activity 'hot-spots', e.g. car parks, footpaths and visitor centres.

▶ There is an observable distance decay of impact away from activity hot-spots.

Careful consideration of these assumptions should be used to assist in the design of a sampling frame and strategy, together with key questions or hypotheses.

Basic survey techniques

To start with, obtain a large scale base map and identify likely areas or zones of recreational pressure. It is a good idea to plan data collection points with increasing distance away from the activity focus. There are a number of options for data collection as outlined below.

1. Determine the path width, morphology, soil depth, infiltration and degree of compaction (supported with annotated field sketches and photographs). There are a number of factors which predispose a footpath to erosion. These can be split into two main categories: physical and human.

Physical factors	Human factors
Angle of slope	Visitor pressure
Soil type and depth	Type and duration of activity
Drainage	Proximity to car park and other facilities
Compaction	Popularity of route/people flows
Climate	Seasonal differences in usage
Vegetation type	Accessibility – signage and proximity to centres of population
Length of growing season	
Altitude and aspect of slope	

Measure the widths of footpaths leading from a car park at several sites. Record evidence of footpath degradation with photographs and note the aspect and slope angle. If paths show signs of gullying it is feasible to measure the degree of path erosion and deformation. Place a taut tape across the path surface to form a small transect line and measure the distance between the tape

and the ground (Figure 2.101). At each site also record plant height, percentage vegetation cover, soil depth and compaction.

2. Use a vegetation transect to determine the percentage cover of plants, plant heights and degree of damage. The vegetation can be displayed as a kite diagram (Figure 2.103) and a semi-quantitative score developed for ecological impact (see page 123).

3. Design a visitor and activity survey to establish the number of people using paths: weekend v weekday, morning v evening.

Also consider making an activity/impact matrix for an area (Table 2.22). This subjectively assesses how the activities interact with each other, the local community and the environment.

Figure 2.102 shows some examples of plants and their sensitivity to trampling. Do your results fit the predicted pattern?

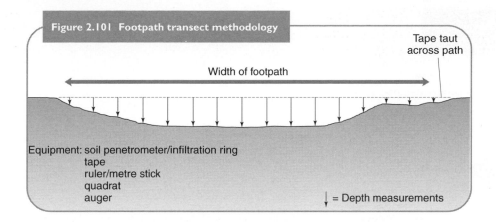

Figure 2.101 Footpath transect methodology

Width of footpath

Tape taut across path

Equipment: soil penetrometer/infiltration ring
tape
ruler/metre stick
quadrat
auger

↓ = Depth measurements

TOP TIPS

Using a simple 1–5 scoring system map areas according to degree of trampling evident (where 1 = no trampling and 5 = serious evidence of trampling). Present your results as a shaded chloropleth map. Use this as a tool to work out where to survey, to ensure appropriate balance and range for your project aims and hypotheses, based on the degree of trampling in an area.

4. A car park survey at regular intervals during the day. Which is the most popular car park and why? Determine the origin of the cars from garage stickers, tax discs and number plates.

5. An environmental quality assessment, including litter survey.

6. A questionnaire survey to determine the sphere of influence and socio-economic characteristics of users.

Plant Name
Ribwort Plantain
Stagshorn Plantain
Daisy
Meadow Grass
Ladies Bedstraw
Clover
Sand Sedge
Common Dandelion
Dragonstail Dandelion
Buttercup
Birdsfoot Trefoil
Salad Burnett
Ragwort
Catsear
Common Storksbill
Lesser Hawkbit
Sea Spurge
Sea Bindweed
Sea Couch Grass
Sow Thistle
Marram Grass
Key
Very Resistant
Resistant
Less Resistant
Mildly Susceptible
Very Susceptible

Most Resistant ↑

Least Resistant ↓

Figure 2.102 Plant susceptibility to trampling

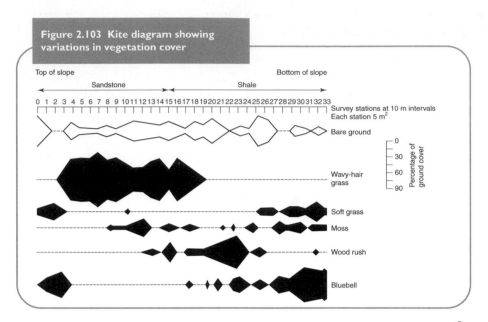

Figure 2.103 Kite diagram showing variations in vegetation cover

A coastal resort impact matrix									Table 2.22
Activity ✓ = peaceful co-existence ? = uncertain ✗ = negative mix	Dog walking	Cycling	Angling	Birdwatchers	Hangliders	Fell runners	Skateboard/rollerblade	Environment	Community
Dog walking	–	✗	?	✗	?	✓	✗	?	✓
Cycling	✗	–	✓	✗	✓	✓	✓	✓	✓
Anglers	?	✓	–	✓	✓	✓	?	?	✓
Birdwatchers	✗	✗	✓	–	?	✓	✗	✓	✓
Hangliders	✓	✓	✓	?	–	✓	✓	✓	?
Fell runners	✓	✗	✓	✓	✓	–	?	✓	✓
Skateboard/rollerblade	✗	✓	?	✗	✓	?	–	?	?

TOP TIPS

There are a variety of cartographical techniques that can be used to display your data. For example:

▶ Isoline map to show people activity or sphere of influence.

▶ Scatter graph of litter count/assessment and distance from hot-spots. If there is a large enough dataset (10 or more pairs) use Spearman's Rank to test for any significant correlations.

▶ Use simple bar graphs to show number of cars at different times of the day or week.

Possible study	Data collection and presentation ideas
1. An account of soil and vegetation variations from the top of a slope to the bottom	▶ Use an interrupted or continuous transect as the sampling frame. ▶ At suitable intervals record soil depth (possibly texture colour etc.) and also monitor changes in plant abundance and type. Gradient can also be measured. ▶ Plot results as a mixture of bars for soil and kites for changes in vegetation cover.
2. How much tree height/girth/density /productivity is controlled by soil depth?	▶ Establish a large 30 × 30m quadrat and survey all the individual trees within it, noting species, girth and height. ▶ Take a number of random soil samples within the sample frame and measure soil depth with an auger. ▶ Repeat the measurements at a comparative site and statistically test for differences between the two datasets.
3. Do urban areas exhibit greater plant biodiversity than the surrounding countryside?	▶ Be careful to establish similar sites for making comparisons, e.g. town park v village green. ▶ Conduct a survey to complete a full species list and establish diversity. ▶ Use Simpson's Diversity Index to quantify results.
4. An evaluation of visitor impact at sand dune (honeypot site) X	▶ Conduct a vegetation survey to establish levels of trampling with distance away from activity hotspots. ▶ Use soil surveying techniques to determine levels of compaction and path morphology. ▶ Conduct a visitor survey and consider determining spheres of influence or constructing an impact matrix. ▶ Record frequency/origin and duration of visitor stay.

SAFETY

▶ When handling and working with soils it is advisable to wear plastic/rubber gloves at all times to minimise risk from toxicara (a disease present in dog faeces). Always wash your hands on return from the field.

▶ If sampling vegetation on slopes take care when ascending and descending steep gradients. Under no circumstances run down hills (no matter how much fun it may seem!).

▶ Remember that grass and other vegetation becomes slippery when wet; wear appropriate footware and take extra care in wet conditions.

WEBSITES

The Soil Survey: www.silsoe.cranfield.ac.uk/sslrc/

www.sandsoftime.hope.ac.uk – details on the Sefton coast sand dunes

3 HUMAN ENVIRONMENTS

RURAL INVESTIGATIONS

RURAL LAND USE SURVEYS
Choosing a study area
Equipment
Carrying out the rural transect
or rural land use samples
Presenting and interpreting
the results
Taking it further

FARM SURVEYS
Ideas for farm investigations
Writing up your farm surveys

INVESTIGATING RURAL SETTLEMENTS
Getting started
The basic building blocks of
village surveys
Carrying out a village
questionnaire
Taking it further

URBAN ENVIRONMENTS

THE CBD
Basic survey methods

SHOPPING STUDIES
Shopping hierarchies
Distribution of shopping types

INVESTIGATING RESIDENTIAL LAND USE
Getting started
Basic residential surveys

Possible investigations within
residential areas

URBAN CATCHMENT AREA OR HINTERLAND SURVEYS
Writing up

POPULATION AND PEOPLE SURVEYS
Example studies

POLLUTION OF THE ENVIRONMENT

INVESTIGATING AIR POLLUTION
Primary data collection
Secondary data

INVESTIGATING NOISE POLLUTION

LAND BASED POLLUTION
Litter surveys
Derelict land and buildings
surveys

TRANSPORT INVESTIGATIONS

TRAFFIC FLOW SURVEYS
Basic surveys
Additional surveys

PUBLIC TRANSPORT SURVEYS
Changes in public transport
networks

ROUTE QUALITY SURVEYS

PARKING SURVEYS

INVESTIGATING THE IMPACT OF TRANSPORT TERMINALS

SPORT, LEISURE, RECREATION AND TOURISM

BASIC TECHNIQUES

Assessment of the supply of provision

Assessment of the demand for particular facilities

Assessment of the impact of a proposed/existing development on the surrounding environment

EXAMPLE PROJECTS

Impact of a football stadium

Golf course development

Impact of sports/leisure centre on the surrounding area

Surveys of parks and open spaces

Recreational usage and potential conflicts at a particular site (e.g. reservoir, beach, national park honey pot)

Impact of tourism at a honey pot site

Allotment surveys

Urban tourism

The impact of tourism on the layout of facilities in a town

3

HUMAN ENVIRONMENTS

RURAL INVESTIGATIONS

Human environments

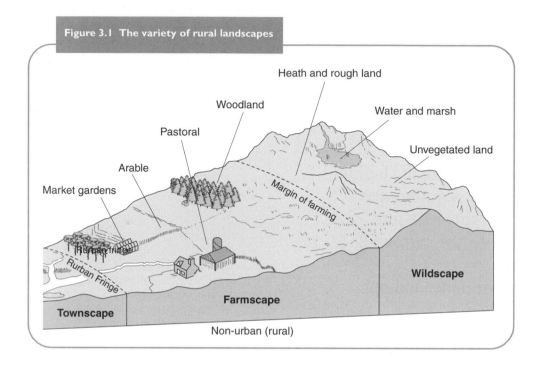

Figure 3.1 The variety of rural landscapes

Rural landscapes cover an enormous variety (Figure 3.1). A basic distinction needs to be made between the wildscape or remote rural areas, in which many of Britain's National Parks are found, and the farmscape, which consists largely of farms and villages. At the fringe of most urban areas is a **rurban** area known as the rural urban fringe which is usually a zone of rapid change as the area is under threat from increased urbanisation.

As all rural areas tend to have lower densities of population, less facilities, and are less built up, one prime consideration is **logistics**. In particular **transport** to and from the area is likely to be infrequent. Although rural areas are some of the most crime free in Britain **never** work alone, even in farmscape areas. The **wildscape** areas are often remote and rugged and therefore having the correct equipment and being with someone who is qualified or experienced in working in hill country is absolutely vital. For this reason, you will usually undertake wildscape fieldwork as part of

an organised group visit often as part of a field trip. One of the key factors to consider is the scope of the survey. If you are working as an individual, a narrow focus on a single farm or village would be appropriate, but if you are working as part of an organised group, some of the more wide ranging surveys, such as the following may be appropriate:

▶ Farm diversification in a valley.

▶ Change in a wide area of rural urban fringe.

▶ Changing services provision in a group of villages.

▶ Establishing a rural settlement hierarchy.

As most land in rural areas is private property, you will need to map out a feasible route using public footpaths for a land use transect, or seek permission from the farmer if you want to carry out a farm survey.

RURAL LAND USE SURVEYS

Choosing a study area

▶ You need an area with a variety of land use, preferably including some arable, some pasture and rough pasture, and a variety of land types (river floodplain etc.). Ideal locations therefore include escarpments or valleys which have variations in slope, aspect and altitude.

▶ Cover a wide enough area to make the results worthwhile but at the same time manageable in terms of fieldwork time. A rural parish is often a good unit because the MAFF statistics (now DEFRA) are published, at this scale.

As you cannot always cover the whole area you will need to **sample** it, either by a **rural transect** surveying a belt of land on either side of a public footpath/country lane (Figure 3.2), or by using **point samples** if you need to look at types of land use related to a particular type of geology or variations in altitude. Expect to sample from an area of around 25 sq km. One problem you will have is finding your sample points in the field – you need to match up the shape of the fields on the map to the field boundaries on the ground.

The best **time** to carry out a rural land use survey is from late spring to early autumn as at that time it is easier to recognise the crops, and the animals will be out in the fields.

Equipment

▶ A base map drawn or photocopied from an OS map at 1:10,000 scale. It must show field boundaries.

▶ A land use key so that you can annotate each field, and a prepared booking sheet for point samples (Table 3.1).

▶ Binoculars for looking at crops/counting stock far away or where access is not possible.

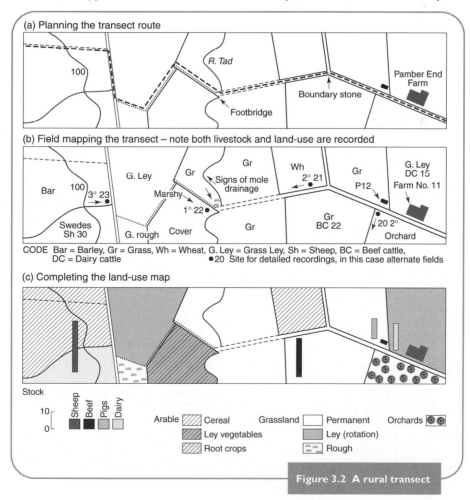

Figure 3.2 A rural transect

Sample booking sheet for part sampling of land use Table 3.1

Sample point G. Ref.	Field size in hectares (measured from map)	Land use in sample field	No. of stock in field	Altitude (from map)	Geology (from map)	Aspect (compass)	Slope (clinometer)	Soil depth in cm	Soil sample results of laboratory tests			Distance from farm	Other comments
									pH	Texture	Moisture content		
1 737204	2.3 ha	Perm. pasture	20 cows	106m	Boulder clay	273° SW	3°	80cm	6.5	Clay	90%	½ km	Field drains

For detailed surveys in sample fields you will need the following.

▶ Clinometer to measure angle of slope.

▶ Auger to measure soil depth.

▶ Compass to record aspect of slopes.

▶ Polythene bags with ties and labels for collecting soil samples for testing in the laboratory .

Carrying out the rural transect or rural land use samples

▶ Record the land use and stocking numbers using your prepared booking sheet.

▶ Record the nature of the fence/hedgerow.

▶ At your chosen sample sites take the necessary measurements. You may decide to do alternate fields on either side of the transect line or alternatively the fields identified by your stratified sample.

The basic aim of a land use investigation is to find out to what extent land use variations can be explained by physical features such as the nature of the soil, altitude or slope. These physical characteristics can be combined to make an assessment of the potential of the land for farming known as **land use capability**. Table 3.2 shows how each factor sampled can be given a class score.

Land use capability Table 3.2

Class	Altitude (m)	Wetness	Soil quality	Soil fertility (ph)	Slope (°)
1 High quality	Below 100	No limitations Free drainage Rainfall <750 mm	Deep soil 75 cm + Stone free Loam texture	7 + (neutral)	Level (not above 3)
2	100–150	Imperfectly drained Drainage easily modified by liming	Depth 50–75 cm Slightly stony	6.0–6.5	Slight (not above 7)
3	150–200	Some problems but possible to install drainage system	Depth 25–50 cm Stony – may be sandy or clayey texture	5.5–6.0	Moderate (not above 11)
4	200–350	Poorly drained but can be improved to maintain pasture	Shallow – under 25 cm Very stony	5.0	Significant (11–20)
5 Low quality	Above 350	Poorly drained Drainage almost impossible to install Rainfall > 1,250 mm	No humus Very stony – skeletal soil only	Under 4.5	Steep (over 20)

Presenting and interpreting the results

▶ The first stage in presenting your results is to complete your land use map. Figure 3.3 shows you the standard key to use (you can then compare your results with any official land use survey).

▶ You then need to complete any necessary laboratory work and calculate your land use capability index. (See p. 134)

▶ Figure 3.4 shows a variety of techniques you could use to analyse the impact of particular physical factors on land use.

Taking it further

If you have time you could follow up your land use transect by arranging to visit the farms involved in your study area/transect to discuss how important the farmer considers physical factors to be and what other factors may have influenced the land use choice.

Other ways you can take your study further include the following.

▶ Comparing your results with historic land use maps in order to assess the degree of change (particularly useful in a rural urban fringe site). To analyse the change from maps use a grid to work out the percentage changes of each land use type.

Simplified land utilisation key (colours in brackets)

Arable – cereals **(yellow)**
Deciduous or mixed woodlands **(green)**
Grassland pasture **(light green)**
Heath, moorland and rough grazing **(yellow/orange)**
Commercial and residential **(grey)**
Industrial land, including utilities **(deep orange)**
Derelict and waste land **(deep grey)**

Arable – market vegetables (including root crops) **(pale yellow)**
Coniferous plantations **(dark green)**
Market gardens, orchards, nurseries, allotments **(purple)**
Open space – parklands, greens, public gardens **(mid green)**
Roads, car parks, airfields, port areas, etc. **(red)**
Quarries, mines and tips **(dark brown)**
Water features, e.g. rivers, lakes, bogs, canals **(blue)**

Note: You can add detail to these 14 basic land use categories by adding letter or number codings to this key, | IN | | SA |
e.g. IN: industrial crops (oil-seed rape), SA: set-aside land. For a more detailed key consult Land Use Maps.

Figure 3.3 Standard key for land use

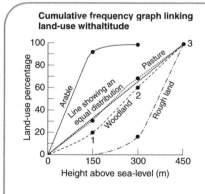

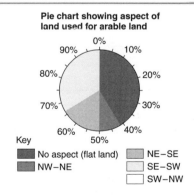

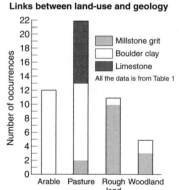

To draw a scattergram you need to know the slope angle of the individual points, rather than grouping the results into broad classes. You mark on each result with a dot. Draw in the median, which is the middle dot of the five for woodland. Note that this is not the same as the mean, which is the average of all the results. You can then divide the recordings into two further halves, so that the results fall into four equal parts. These new lines are known as quartiles, because they divide the list into quarters. The upper quartile and lower quartile will enclose 50% of all the points. The difference between the upper quartile and lower quartile values is

$$\frac{IQR}{Median} \times 100\%$$

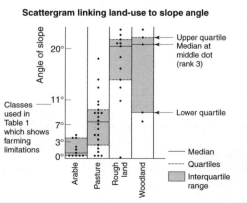

The technique of drawing a cumulative frequency graph can be shown by looking at the woodland line. One fifth of the woodland is found below 150 m, so a dot is shown at point 1, showing that 20% of the woodland is found below 150 m. It can be seen that a further 40% of the woodland is found between 150 and 300 m. However, to find the cumulative amount of woodland below 300 m, add the two amounts together (20% + 40% = 60%). Therefore you put your second dot at point 2, showing that 60% of the woodland is found below 300 m. Point 3 records the fact that all (100%) of the woodland is found below 450 m. The points should then be joined up with a smooth curve.

Figure 3.4 Presenting the results of a rural transect

▶ Using the MAFF (now DEFRA) statistics, both present day and historic, record how land use has changed over time. These statistics also give details of farm size, ownership, stocking rates and crop hectarage by **parish**.

Current statistics can be obtained from Guildford www.maff.gov.uk. Historic parish records from 1866 of agricultural returns can be read at the Public Record Office in Kew www.pro.gov.uk.

FARM SURVEYS

A farm study represents an ideal option for those students who live in rural areas and have a farming background. In order to get precise information, you need to be certain of the co-operation of the farmer(s) before starting the project. Farmers tend to be very busy especially at harvest time and if you are not known at the farm it is very important that you make an appointment before starting on your project.

A Farm Questionnaire

Name of farm _____   Grid reference _____

Size of farm _____   Altitude _____

Type of tenure _____   Age (if known) _____

Parish _____

LABOUR   No. of household members. Full _____ part-time _____ seasonal _____

No. of outside employees. Full _____ part-time _____ seasonal _____

Specific skilled workforce (e.g. herdsperson) _____

MACHINERY in ranked order by purchase cost _____

LAND USE Arable _____ %, Pasture _____ %, Woodland _____ %, Rough pasture _____ %,

LIVESTOCK	Type	Numbers	Time of year bought/where	Time sold/where	Expected life cycle
CROPS	Type	Acreage	Yield per unit area	Harvest time	Main use/where sold

FERTILISERS   Type and amount per year _____

STOCK FEED   Type and amount per year _____

FUEL   Amount per year _____

PESTICIDES & HERBICIDES   Type and amount per year _____

SUBSIDIES   Name and type _____   Approx. % of total income _____

SPECIFIC PROBLEMS OF FARM _____

_____

_____

PATTERN OF LAND USE Explain why the farm is organised as it is

_____

_____

THE FARMING YEAR Describe the main activities by months

_____

_____

Figure 3.5a A farm questionnaire and a farming system diagram

Farming System Diagram for Stoneybank Farm (A sheep farm in the Welsh Borderland)

Economic Inputs (variable)

Transport
- 4 wheel drive truck and sheep box
- Hauliers

Markets
- Sold to other farmers
- Abattoirs

Capital
- Capital investment (including loans) is 10% of approximate annual costs

Technology
- (Ranked in purchase cost from highest to lowest)
- Truck
- Sheep box
- Tractor
- Fertiliser spreader
- Topper
- Electric fencer

Buildings
- Present lambing barn on site was previously an aircraft hangar

Government
- Subsidised (subsidy = 20% of annual income)

Labour
- Farmer (sheep and farm management) – Mr Dallyn – full time
- Wife (accounting and assistance) – full time
- Casual labourer – full time
- Assistants during lambing season – one person day shift and one person night shift
- Contractors for hedging and cultivation

Behavioural elements

Age
- 45–50

Ambition
- To support family and maintain current success

Knowledge + Experience
- Background knowledge of sheep

Physical inputs (natural)

- Temperature 2–18°C
- Altitude 93 mamsl
- Slope 37°–45°
- Aspect North Easterly
- Soils pH 5.5

Cultural inputs (human)

Tenure
- Owned outright by Mr Dallyn

Farm Size
- 205 acres

Decision making

- Possible uses of land
- Steep nature of land removed possibility of dairy or arable farming
- Mr Dallyn had background knowledge of sheep
- Flexibility of the sheep system and low capital required was influential

Farm Processes and Patterns

Sheep Farming

- 600 sheep (420 Mule × Texel, 380 Mule × Suffolk)
- 3 sheep per acre
- 5 year life cycle
- Sheep are put onto cattle/dairy farms from November to March
- All fields are used for grazing from April to September
- During lambing season sheep are brought closer to farm buildings
- Sheep are regularly vaccinated using drenches and are dipped in organphosphorous at least twice a year
- Also vaccinations against common diseases
- Regular accounts are kept for National Sheep Organisation

Outputs

From Processes

- Spent dip waste products must be neutralised (adding to costs)

From Sheep

- Meat and lambs (to other farmers)
- Wool (secondary purpose)

Feedback

- Satisfactory income from sheep sales and subsidy

Breakdown of sources of profit from farming processes

Sheep sales 80% | Subsidy 20%

- Farming is however not the main source of income for the farmer
- Main source of income is from contracting business run alongside the farm

Figure 3.5b A farm questionnaire and a farming system diagram

Figure 3.5a shows a farm questionnaire based on inputs, process and outputs; Fig 3.5b shows the sort of summary diagram you could develop from the questionnaire. Issues of cost are often sensitive to farmers, so be prepared to be flexible and talk about **relative costs**. If you need advice as to choice of farm consult your local branch of the National Farmers Union (NFU) or the Country Landowners Association (CLA).

Ideas for farm investigations

1. A study of a **single farm unit** to see to what extent the pattern of farming is influenced by physical, socio-cultural or economic factors. If the farm is comparatively small (under 50 hectares or 125 acres), it is best to compare two farms of a similar size or on similar sites.

2. A study of a single farm to see how and why land use has changed over a particular period of **time** (Figure 3.6). For this type of study ensure the farmer has sets of farm records and maps dating back at least 30 years. You can study the impact of technology and farming subsidies and compare the changing inputs and outputs.

3. A more general farming survey (best done by a group) involving the study of around 10–20 farms within a defined area (perhaps three parishes or a valley) to investigate a particular issue. You could look at one or more of the following.

▶ The variations in land use with reference to farm size and quality.

▶ The impact of farmer's values on farming decisions; farmers farm similar land areas in different ways often based on personal preference.

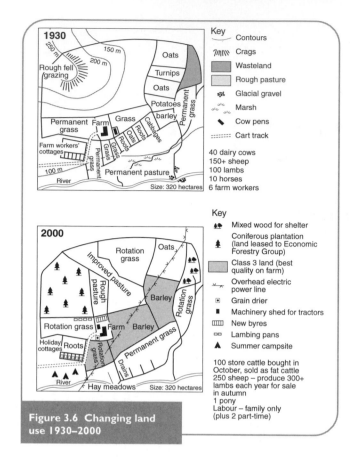

Figure 3.6 Changing land use 1930–2000

▶ The impact of government policies such as milk quotas, diversification grants, the removal of hedgerows or the impact of new conservation grants such as organic grants.

▶ The impact of a particular event, e.g. restocking after foot and mouth disease.

▶ The problems and concerns of the rural-urban fringe.

If you are doing comparative studies it is always worth arranging a meeting with the NFU (local office) or the Chair of the CLA, or visiting a

								Soil sample analysis								
Field number, which should be recorded on your map	Size	Crop	Livestock type and numbers	Altitude of centre	Slope angle	Soil depth	pH	Texture	Moisture content	Nitrogen content	Fertiliser used	Drainage systems	Boundary	Distance from farmhouse	Land capability rating
1	3 ha	Wheat	–	102 m	3°	85 cm	7	Loam	70%	80%	Phosphates Nitrates	Mole drains	Hedge-row	1.4 km	1.3

Farmer's field booking sheet **Table 3.3**

meeting of the Young Farmers to explain your ideas and to find possible farms to survey.

4. You could look at farm diversification. A project of this type would consist of two major parts.

▶ The **questionnaire** discussion with the farmers/farm managers.

▶ The **farm survey**. Come prepared with a 1:10,000 map so you can talk to the farmer and put on the farm boundaries, although many farmers do have farm maps. When you go round the farm use a booking sheet to record details of each field (Table 3.3).

Additionally it is a good idea to survey the farm buildings and to make a sketch plan showing what each is used for. Find out about size, use, age. It is very sensible to take photographs of the general site, situation and any interesting features such as new cattle pens, or conservation measures.

TOP TIPS

Keep the aim of your investigation firmly in mind as it will affect how you structure your work. It is easy to be descriptive and provide endless pictures of farm machinery.

Writing up your farm surveys

You should aim to include the following when writing up farm surveys.

▶ Detailed annotated maps of land use, stocking rates, and how physical factors affect land use. You can also link land use intensity to distance from the farmhouse.

▶ A sketch plan/map of the main use of all of the buildings.

▶ A land capability matrix and map to show variations on the farm.

▶ Stocking numbers throughout the year and percentage land use of each crop with a diagram of any crop rotations.

▶ A graph of worker hours per year (you can calculate this using Table 3.4). For example, Table 3.4 shows that if you have 10 hectares of oilseed rape it will take around 200 worker hours per year.

If you are looking at **change**, this check list will need to be developed for key dates, for example when the farmer made significant changes, so that you can compare changing yields per hectare, per animal, amounts of machinery, workers employed, carrying capacities of stock per hectare and changing building use.

For more general farm surveys you need to ensure that you present comparative data effectively, on tables and maps, and then try to explore the key factors. For example when looking at comparative land use, you will need to consider type of land, farm size, or distance form nearest markets. Figure 3.7 shows how you might investigate the significance of the personal decisions of the farmer. You can relate the responses to the following.

▶ The nature of the farming area.

▶ The type of farm.

▶ The farmer's age and educational background.

▶ The type of tenure.

▶ Whether the farmer comes from a traditional farming background.

Labour requirements for crops and stock		Table 3.4

CROPS (worker-hrs per ha. per year)		Stock (w-hrs per head per year)	
wheat, barley, oats, rye	20	beef cattle	22
oilseed rape	20	dairy cows	55
cabbage, caulis, sprouts	350	pigs	29
apples, pears	550	goats	55
strawberries	1400	sheep	4
blackcurrants	1000	laying	0.2
gooseberries	920	broiler chicks	0.02
bulb flowers	1600	geese	0.2
heated greenhouse	18,000	turkeys, ducks	0.08
non-h. greenhouse	14,000	rabbits	0.07
turnips, swedes	440		
carrots	310		
potatoes	240		
permanent grass	9		
rough grass	2		

The farmer's values (score 0 if irrelevant, score 3 if essential, score 1 or 2 if intermediate)

Healthy outdoor lifestyle	☐
Maintaining the farm traditions	☐
Independence	☐
Modernisation (technical improvements)	☐
Maximising income	☐
Working with animals	☐
Changing occupation	☐
Able to influence farming policies and market prices	☐
Flexible working hours	☐
Passing the farm into the family	☐
Satisfaction of producing healthy food	☐
The weather	☐
Being an environmental caretaker	☐
EU subsidy maximisation	☐
Producing high quality products	☐
Desire to innovate	☐
Others	☐

Figure 3.7 Factors influencing farming decisions

INVESTIGATING RURAL SETTLEMENTS

The main unit of rural settlement is the village which can be defined as 'an agglomeration of houses, which provides the basic services for the community'. There are an enormous range of project titles or investigations which are possible. They fall into two groups.

▶ Single or two village studies which are suitable for individual small scale investigations. When comparing two villages always be sure to pick two villages which have something in common, e.g. size, position or function.

▶ Looking at a village's relationship with the wider area, for example as part of a settlement hierarchy survey. This type of project is particularly suitable for groups of students, working as a team they can cover a whole range of settlements within a designated area, and then individual titles can be generated within the group work. These type of studies are called multiple village studies.

The basics of a village survey **Table 3.5**

	Definition	What to look for	Sources
Site	The land the village is built on	Aspect – which way the settlement faces Relief – whether land is sloping or flat Altitude – how high above sea level Geology – whether firm or well drained Water – whether free of flooding or water supply Soil – whether fertile for farming Shelter – from the wind Space – whether room for expansion	From 1:25,000 map From 1:25,000 map From 1:25,000 map From geology map From 1:25,000 map From soil sample and pH From field survey and 1:25,000 map From field survey and 1:25,000 map
Situation	The position of the village with reference to the surrounding area	Position with reference to major roads, railways and nearby towns	From 1:50,000 map
Form	The shape of the village	The shape of a village can be classified whether it is linear or round, or compact or spread out, or is formed around a green	From 1:10,000 map and field survey
Function	What the village does	Employment surveys of the villagers Population surveys to see the age structure of village Building use survey of housing and services	Questionnaire Questionnaire Field map
Growth	How the village has changed over time	Growth of the population Growth of the settlement Growth of the services	Census totals over time Comparative historic maps at 1:10,000 scale Field map of age of the buildings

Getting started

Making a basic village survey is necessary for all types of investigation. Table 3.5 summarises the key data which need to be collected for a village survey. The basic village survey involves several stages.

1. Obtaining the relevant large scale maps, and background data, such as census figures, and preparing appropriate booking sheets. Local guides can be very useful but it is easy to be led away from the essential geography.

2. Using OS maps to carry out a **site**, **situation** and **shape** analysis. www.multimap.com

3. Mapping the main features of the village by walking through the village. Without **an age and function survey** most investigations will prove difficult. Try to concentrate on as many types of survey which will not require questionnaire work, for example:

   ◗ housing condition

   ◗ house price

   ◗ environmental quality

   ◗ traffic flow and speed

   ◗ bus services

   ◗ litter and noise surveys.

4. **Questionnaire work** is sometimes difficult to undertake in a village unless you live there and know all the people. You could make some prior arrangements to go to the village school and work with the children or visit the local Women's Institute in session. When you go to the village you may find it to be very quiet, and except for the occasional visitor to the village shop or post office there is nobody to question. In general knocking on unknown people's doors can be difficult so you need to be prepared for a limited response to any questionnaire and also for the fact that it will provide an unrepresentative sample. During the day villages may be full of elderly people, or young mothers who are not at work.

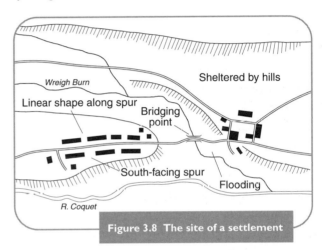

Figure 3.8 The site of a settlement

A site evaluation table   **Table 3.6**

	V. good ←					**→ V. poor**	**Village name** _____	
Feature	**+3**	**+2**	**+1**	**0**	**–1**	**–2**	**–3**	**Guidelines**	**Method**
Aspect		✓						Faced SE	Map and compass
Height	✓							30 m	Map
Shelter		✓						Good except for SE winds	Field
Freedom from floods	✓							On chalk No danger of stream flooding	Map and field
Access to water	✓							Spring in village	Map
Firmness of ground		✓						Subsidence only on very steep slopes	Geology map and field
Room for expansion			✓					Limited to N and W	Map and field
Fertility of soil		✓						pH8	Field tests
Defensive quality						✓		Poor, especially from West	Map and field
Other	✓							Bridging point	Map and field
	12	8	1		–1				

Total =   20 (21 – 1)

The basic building blocks of village surveys

The **site** of a village is the land the village is built on and can best be shown on a sketch map or using a photograph and annotating it (Figure 3.8). Go up to a high view point to take the photograph. The site includes relief, altitude, aspect, shelter, water supply, drainage, soil, firmness of ground, geology, and space, i.e. whether there is room for expansion. A site evaluation index such as that shown in Table 3.6 will give you an idea how favourable the site is. You can then compare it with other neighbouring villages.

The **situation** of a village is its position with reference to the surrounding area. A **nodality index** (Table 3.7) gives an indication as to how important a village is with reference to the surrounding area. Most of the information can be obtained from the 1:50,000 OS map for this index.

A favourable site combined with an important situation should encourage a village to develop – but this is not always the case.

The **shape** of a village is its form. Figure 3.9 shows you how to classify village shapes. The shape is a result of how the village has developed.

The **growth** of a settlement or its decline is a complex concept. Population growth can be analysed using a series of census records from 1801, (Figure 3.10) but there are a number of problems such as boundary changes or the fact that the figures are for a whole parish – not just the village. You can **estimate** the population by counting the number of houses and using the family multiplier (2.6 for rural areas) but beware of second homes. In some Lake District villages only 15 per cent of the vilagers are resident population. Alternatively you can build up a population pyramid by direct fieldwork (Figure 3.11).

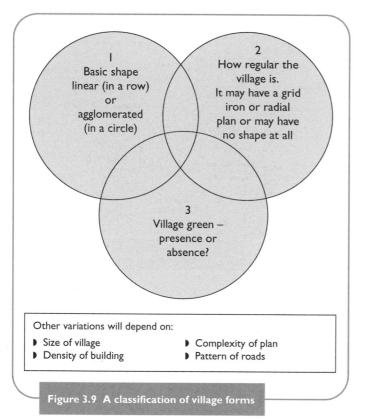

Other variations will depend on:

▶ Size of village   ▶ Complexity of plan
▶ Density of building   ▶ Pattern of roads

Figure 3.9 A classification of village forms

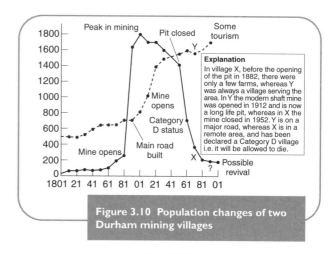

Figure 3.10 Population changes of two Durham mining villages

Village	Population from last census	Service count from O.S. map and by fieldwork	Route count 'A' road = 3 'B' road = 2 Other = 1	Nodality Index
Fernham	870	|||	2 × 2 = 4 2 × 1 = 2	9
Stanford	1,320	ﬀ|	2 × 3 = 6 0 × 2 = 0 1 × 1 = 1	13
Compton	340			3 × 1 = 3	4

Calculating a nodality index   **Table 3.7**

The **growth** of a village can best be seen by looking at the increase in the area of land it covers. You can build up an age of housing map by using a combination of old maps and visual surveys (Figure 3.12). Some houses have their date of building on them. Building material and house type surveys can also be carried out as you make your age of building map.

The **function** of each building shows what people do in the village. A building's function consists of two aspects.

▶ What the building is used for (Figure 3.13)

▶ What people do for a living. This can only be found out for the present day by using a village questionnaire, but census enumerater's are now available up to 1901, so can provide a historic contrast. Write to your local public records office

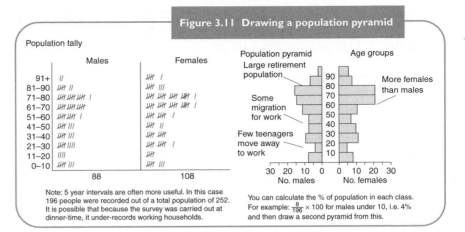

Figure 3.11 Drawing a population pyramid

Population tally

	Males	Females
91+	//	
81–90	ЖĦ //	ЖĦ ///
71–80	ЖĦ ЖĦ ЖĦ /	ЖĦ ЖĦ ЖĦ ЖĦ /
61–70	ЖĦ ЖĦ ЖĦ	ЖĦ ЖĦ ЖĦ ЖĦ /
51–60	ЖĦ ЖĦ /	ЖĦ ЖĦ /
41–50	ЖĦ ///	ЖĦ //
31–40	ЖĦ ///	ЖĦ ЖĦ
21–30	ЖĦ ////	ЖĦ ЖĦ /
11–20	////	ЖĦ
0–10	ЖĦ ///	ЖĦ ///
	88	108

Note: 5 year intervals are often more useful. In this case 196 people were recorded out of a total population of 252. It is possible that because the survey was carried out at dinner-time, it under-records working households.

Population pyramid — Age groups

Large retirement population — Some migration for work — Few teenagers move away to work

More females than males

No. males — No. females

You can calculate the % of population in each class. For example: $\frac{8}{196} \times 100$ for males under 10, i.e. 4% and then draw a second pyramid from this.

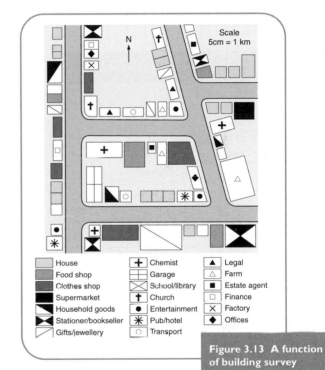

House		Chemist		Legal
Food shop		Garage		Farm
Clothes shop		School/library		Estate agent
Supermarket		Church		Finance
Household goods		Entertainment		Factory
Stationer/bookseller		Pub/hotel		Offices
Gifts/jewellery		Transport		

Figure 3.13 A function of building survey

Figure 3.12 An age of building survey

Map number	Use	House type	Building material	Roof material	Age
1	House	D	Brick	Tile	1960
2	House	D	Brick	Tile	1960
3	House	Cottage	Half Timber	Thatch	1700
4	Stables (former chapel)	–	Brick	Slate	1863
5	House	D	Stone	Slate	1880?
6	House	SD	Brick	Tile	1965
7	House	SD	Brick	Tile	1965

Numbered buildings match up to booking sheet — High Ercall

Rodington Heath — Inn — Green — School — Church — Rectory — Longden — Atcham

Key
■ Pre-1902 buildings

Not to scale

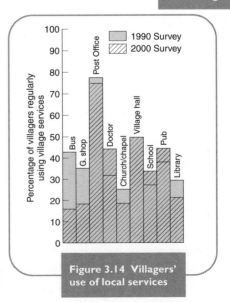

100 90 80 70 60 50 40 30 20 10 0

Percentage of villagers regularly using village services

☐ 1990 Survey
▨ 2000 Survey

Bus — G. shop — Post Office — Doctor — Church/chapel — Village hall — School — Pub — Library

Figure 3.14 Villagers' use of local services

for information. For some parishes, records are available at the church of births, marriages and deaths and they show occupations too.

▶ You can also find out how many people use village services (Figure 3.14). Schools, churches, clubs etc. will have registers, and many village stores will be able to tell you where their customers come from – usually only a very local catchment.

Carrying out a village questionnaire

Figure 3.16 shows a **household questionnaire**. If the coverage is good you can actually make your own

population pyramids to show the structure of the village population. You can also compare any new estates with the rest of the village, by carrying out a sample of 50 from each group.

Assessing rural deprivation is another type of rural village project which depends on questionnaire work. Rural deprivation consists of three facets.

▶ **Household deprivation** is concerned with the hardships of individual families trying to maintain a living standard and can be identified using small area census statistics for a village combined with sensitive deductions from questionnaire work. Key indicators of poverty could include percentage of pensioners, percentage of low paid

---

Figure 3.15 A village survey questionnaire

Village Survey Questionnaire

A. FAMILY STRUCTURE

Male	Female	
☐	2	Pre-school
☐	☐	School age
1	1	Young single/marrieds
☐	☐	Middle-aged
1	☐	Pensioner Total 5

B. HOUSING

Cottage	☐	Pre-Victorian	☐	H & C water	✓	Rented	☐
Terrace	☐	Victorian	☐	Bath	✓	Council	✓
Semi	✓	Inter-war	✓	Inside toilet	✓	Owned	☐
Detached	☐	Modern	☐	Central heating	✗	No. of rooms excl. WC, hall, etc.	6

C. OWNERSHIP

Recent surveys indicate the following are vital for country dwellers: (✓ if you have one)

Car	✓	Freezer	✓
Telephone	✓	TV and video	✓

D. WORK

	Place	Nature of work	Transport
Main income earner	Plotts Farm, Windyhaugh	Farm Work	Walk
Second income earner(s)	Windygyle Lodge	Cleaning	Car

E. SERVICES

Location and means of travel to:

Primary school	Rothbury	Bus	Doctor	Rothbury	Car	Weekly groceries	Alnwick	Bus/car
High school	Morpeth	Bus	Dentist	Rothbury	Car	Meat	Rothbury	Car
Petrol	Netherton	Car	Vet	Alnwick	Car	Entertainment	Newcastle	Bus
Garage repairs	Rothbury	Car	Nurse	Thropton	Bus	Sport	Newcastle	Bus
Solicitor	Rothbury	Car	Hospital	Ashington	Car	Clothes/shoes	Newcastle	Bus

F. PROBLEMS

Choose the three major problems, number in order of importance:

High cost of living	1	Lack of teenage clubs		Lace of suitable jobs	?
Need to own a car	2	High cost of housing		Loneliness for young children	3
Lack of bus service		Lack of middle-age clubs		Others – please state	
Winter snow problems		Problems of health care			

G. MIGRATION

Has any member of your family moved away? Brother Reason Work

Do you want to move away? No Reason _____

TOP TIPS

Make your sample as comprehensive as possible and 'tailor make' the questionnaire to the precise purpose of your study. You can either do a household (door to door) or village questionnaire. Think carefully which is best for your needs and which is feasible. Figure 3.15 shows a booking sheet for a village questionnaire but note if you want to look at how a village is changing with the arrival of new commuters, a household questionnaire would be better. This is best done on a sunny or summer evening or weekend when most people are in their gardens.

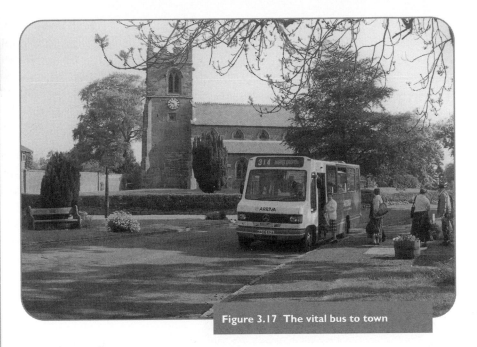

Figure 3.17 The vital bus to town

work such as farm work, percentage of unemployed, and the incidence of households without basic services such as central heating, cars etc.

▶ **Opportunity deprivation** is concerned with the lack of access to services such as hospitals, low cost shopping and petrol and this can be assessed

along with **mobility deprivation**, which is the lack of transport for obtaining basic services.

From the results of the questionnaires, or from general enquiries at the village store and post office, it is also possible to develop an index of accessibility for a particular village which shows how far people have to travel for services. Figure 3.17 shows the importance of the local bus.

So far we have looked at a number of surveys which could be carried out in a village. The challenge is choosing a village you can reach that you know

Figure 3.16 A household questionaire

Sheet	Years lived here	Where did you move from?	How many cars do your own?	Place of work	Job	Where do you shop for: Weekly shop?	Clothes/ shoes?	Name them main advantage of living here	Name a disadvantage of living here	Comments e.g.. on family size, age/ sex, housing type
1	2yrs	London	2	Reading	Programmer	Thatcham	Reading	None	No night life meeting other mums	Detached, 30+ 2 yg children
2	1 yr	Reading	1	Newbury	Teacher	Sainsbury Calcot	Reading	Quiet	No play groups	Detached, 30+ 2 yg children
3	1/2 yr	Guildford	2	Reading	Civil servant	Tesco Chinham	Reading	Nice house	mud	Detached, 25, child on way!
4	2 yrs	Corby	2	Bracknell	Engineer	Tesco Chinham	Reading	Good walks	need a car	Detached, 29, 1 yg child another on way

you will enjoy working in and finding a title which will work.

Taking it further

In order to take your project further you could consider the following ideas.

▶ Investigate how a village has developed over time, and look at the key factors influencing changes in population, employment and services.

▶ Investigate the quality of a village environment in terms of pollution.

▶ For more remote villages you could focus on the issue of deprivation, or discuss to what extent a village is self contained, or has inadequate services.

▶ For some **honey pot villages** you could assess the impact of tourism on the village (services, environment, etc.).

▶ For suburbanised villages it is also possible to carry out an impact survey of the new large estates (traffic flow, household profile, services, etc.).

▶ A particular village can be compared to either a classic model of a village (Figure 3.18a), or to a model of a suburbanised village (Figure 3.18b).

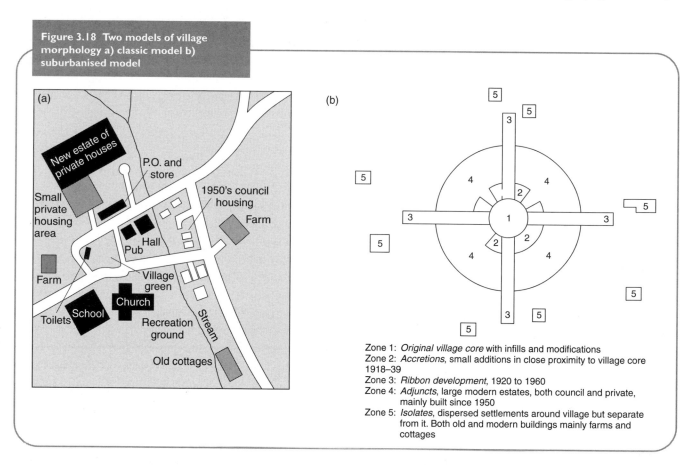

Figure 3.18 Two models of village morphology a) classic model b) suburbanised model

Zone 1: *Original village core* with infills and modifications
Zone 2: *Accretions*, small additions in close proximity to village core 1918–39
Zone 3: *Ribbon development*, 1920 to 1960
Zone 4: *Adjuncts*, large modern estates, both council and private, mainly built since 1950
Zone 5: *Isolates*, dispersed settlements around village but separate from it. Both old and modern buildings mainly farms and cottages

URBAN ENVIRONMENTS

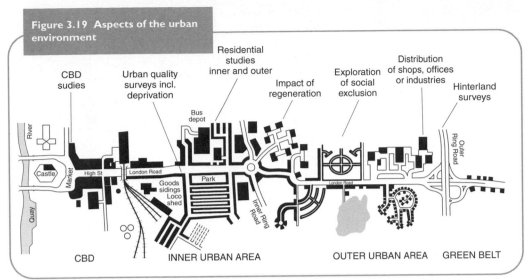

Figure 3.19 Aspects of the urban environment

CBD sudies

Urban quality surveys incl. deprivation

Residential studies inner and outer

Impact of regeneration

Exploration of social exclusion

Distribution of shops, offices or industries

Hinterland surveys

Bus depot

River

Castle

Market

High St

London Road

Goods sidings Loco shed

Park

Inner Ring Road

London Road

Outer Ring Road

Quay

CBD | INNER URBAN AREA | OUTER URBAN AREA | GREEN BELT

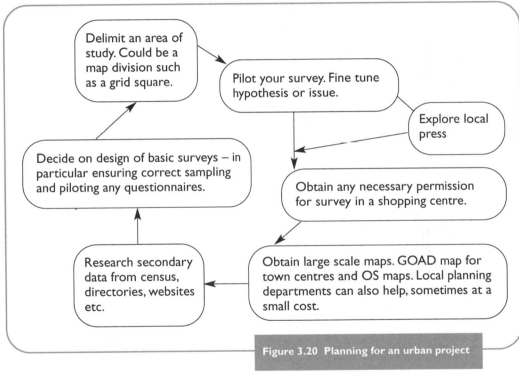

Delimit an area of study. Could be a map division such as a grid square.

Pilot your survey. Fine tune hypothesis or issue.

Explore local press

Decide on design of basic surveys – in particular ensuring correct sampling and piloting any questionnaires.

Obtain any necessary permission for survey in a shopping centre.

Research secondary data from census, directories, websites etc.

Obtain large scale maps. GOAD map for town centres and OS maps. Local planning departments can also help, sometimes at a small cost.

Figure 3.20 Planning for an urban project

Figure 3.19 shows you some of the aspects of an urban environment you could study. You can see there is a huge range. It does not have to be housing or CBD. In Britain around 80per cent of the population live in an enormous variety of urban or built up areas, so if you want a project on your doorstep it is likely to be an urban one. One major advantage is that the geography of urban areas is usually rapidly changing and you can focus on a comparatively small area because of the concentration of geographical activities within it. Further advantages include plentiful supplies of secondary data and with the most recent census in 2001, it will be up to date. Figure 3.20 shows some essential planning and preparations for all urban projects.

THE CBD

The Central Business District (CBD) is the commercial centre of a town. It is characterised by a concentration of shops, offices, public buildings and places of entertainment. As the most accessible part of the town it is the focal point of the transport network (bus station, train station, numerous car parks). Many shops and offices need to be easily accessible to as many people as possible and are therefore prepared to pay high rates to set up in such a desirable town centre site. The most valuable site, which should be the most accessible area, is called the PLVI (Peak Land Value Intersection). In recent years, CBD sites have lost some of their attractions because of congestion, excessive costs, and a shift of the centre of gravity towards the suburbs with the development of out of town centres.

There are a number of basic surveys you can carry out within a CBD. Working as a group you should be able to produce sufficient data in around 3–4 hours. As an individual at least two days' fieldwork (much is dependent on the size of the CBD) would be required. For some surveys such as **standing pedestrian counts** which need to be carried out at a range of locations

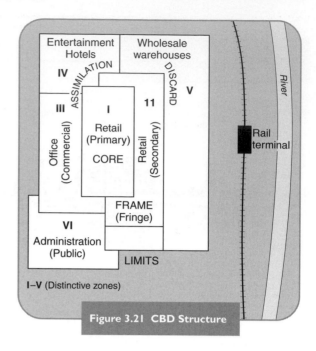

Figure 3.21 CBD Structure

simultaneously, methods have to be modified. Figure 3.21 shows the essential features of the **core** or heart of the CBD and the **frame** of the area surrounding it. At the CBD **fringe** you should be able to see a distinct change from the commercial character of the centre. Some ideas for projects are outlined below.

Basic survey methods

Land use survey

The aim of a land use survey is to find out what each building and open space is used for and to record it on a large scale map. An ideal base map which shows each building clearly will either be an OS map at a scale of 1:2500 or a GOAD map (obtainable from GOAD map shop, 8–12, Salisbury Square, Old Hatfield, Herts, AL9 5BJ).

When you are conducting fieldwork, you will find it easier to use a code on your map. Table 3.8 shows an example of a standard yet simple land use classification which works well for a CBD.

If you are working as part of a group you will have an assigned area with named streets. Walk through the town recording on your map the ground floor use of every building (Figure 3.22). Land use for the upper stories should also be recorded on a separate base map. Figure 3.23 shows you how to convert your field map to a finished land use map.

Once you have a complete land use survey you can use it for a number of purposes.

- Producing an overall shop count (use a bar graph to show the overall shopping provision and a pie chart to show the percentage of each type).

- Producing transects showing land use changes.

- Producing maps to show the distribution of individual, business or professional services or shop types such as clothing shops or shoe shops. You can then analyse their distribution using nearest neighbour analysis as shown in Figure 3.24. Clustered services often result because of **functional zoning** in a CBD.

- Producing diagrams to show how land use varies with distance from the town centre in 100 or 200 metre intervals.

- You can also produce indexes which calculate the percentage CBD to non CBD use for each street.

- Your land use map combined with a GOAD map is also ideal for producing an **index of shopping health**. This is the proportion of derelict/vacant premises, charity shops and other low value

Possible studies	Possible survey data collection
1. How and why does pedestrian density vary in a town?	Pedestrian surveys linked to shopping quality/street appearance. Shopping index of health and distance from PLVI.
2. How and why does shopping quality vary in a town?	Shopping quality surveys. Shopper's perception surveys. Link to pedestrian counts and distance from PLVI. Index of health surveys.
3. How and why do three designated shop types or services (business and professional) vary in distribution in a town?	Land use map. Concept of clustering for comparison shopping or business services. Relationship to distance from town centre. Use of nearest neighbour analysis.
4. How and why does shopping provision vary in different streets?	Land use surveys. Land use transects. Shopping quality and health, pedestrian counts, footfall surveys, distance from town centre.
5. What is the impact of a particular group on shopping provision (e.g. university students in a university dominated town) or tourists in a tourist dominated CBD?).	Land use survey. Pedestrian counts can be age related or pedestrian type related.
6. How user friendly is a shopping centre for use by disabled people?	Land use map. Survey of disabled facilities by shop, e.g. wheelchair access (list available from most town halls to develop grading scale). Mobility index. Specific disabled facilities, e.g. toilets and car parking within shopping area.
7. How and why does land use change in a transect N–S or W–E in a CBD?	Must be a large CBD as the transect is a sample across the CBD zones. Shopping quality, pedestrian densities, footfall surveys, changing land use, car parking regulations.
8. Delimiting the CBD core and frame.	Pedestrian count, shopping quality score, land use, shopping health etc. to provide a series of CBD limits. You can do this for both core and frame.
9. Assessing the reasons for the size and shape of catchment area.	Use theoretical models to devise possible catchment area. Map actual catchment using desire lines from questionnaire results. Link to shopping quality, diversity, nature of facilities, access, car parking, etc.
10. The impact of market days on a town centre or CBD.	Conduct surveys on market days and non market days (pedestrian counts, car parks, catchment areas, bus loading and litter surveys). Interview local traders to assess impact.

Land use classification for a CBD | Table 3.8

Field Map Symbol	Description	Field Map Symbol	Description
A	**Major shopping units** e.g. department/variety stores	H	**Car sales**
B	**Clothing and shoe shops**	J	**Professional services and offices** e.g. banks, solicitors, architects, doctors, estate agents, opticians, chemists, accountants
C	**Convenience shops** e.g. food, tobacconist, newsagent, sweets		
D	**Furniture and carpets**	K	**Public buildings and offices** e.g. school, library, Town Hall, Government offices, G.P.O., police station, church, Job Centre
E	**Specialist shops** e.g. books, sport, jewellers, electrical, hardware, florist, antiques, etc.		
F	**Personal services** e.g. hairdresser, shoe repairs, dry cleaner, launderette, T.V. rentals, gas/electricity showrooms, travel agents	L	**Transport** e.g. car parks, rail/bus station
		M	**Change** e.g. vacant premises, derelict, under construction
G	**Catering and entertainment** e.g. pubs, cafes, hotels, cinema, etc.	N	**Residential**
		P	**Industrial**

Figure 3.22 Land use map

Key
- Department/variety stores
- Clothing/shoe shops
- Convenience stores
- Furniture/carpets
- Specialist shops
- Personal services
- Catering/entertainment
- Professional services
- Vacant premises

shops in relation to other units. If the space occupied by such land uses is disproportionately large the area is termed a **zone of discard**. You can also calculate an **index of diversity for each street**, or gridded square.

$$\text{Diversity Index} = 1 - \sum (X/N)^2$$
X = number of shops in a particular category
N = total number of shops

Work out X/N for each shop category and add up all of the squared values to give a total. The diversity index ranges from 0 – 0.99, a value nearer one indicating greater diversity. You would expect greater diversity in lower value areas of the CBD.

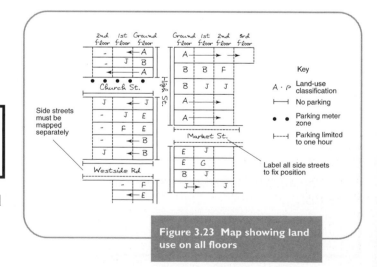

Figure 3.23 Map showing land use on all floors

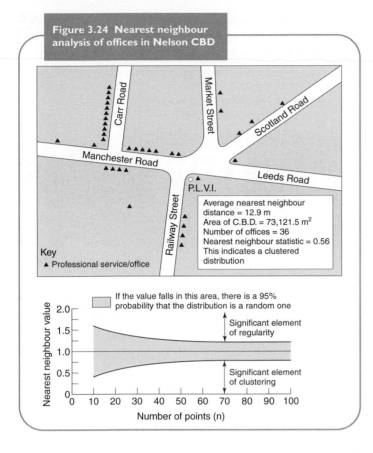

Figure 3.24 Nearest neighbour analysis of offices in Nelson CBD

Average nearest neighbour distance = 12.9 m
Area of C.B.D. = 73,121.5 m²
Number of offices = 36
Nearest neighbour statistic = 0.56
This indicates a clustered distribution

Key
▲ Professional service/office

If the value falls in this area, there is a 95% probability that the distribution is a random one

Significant element of regularity

Significant element of clustering

Around 11.00am and 2.30pm are usually good times. Ideally the counts should be taken at exactly the same time, using the same counting technique, e.g. **one** side of the road, both directions. For busy areas use a mechanical counter. Two minutes is usually enough to produce reliable results, so around four sites can be visited by each group within ten minutes. You can also calculate a measure of crowding by calculating pedestrian density per unit area.

Use an **isoline** map as shown in Figure 3.25 to display this information. As you can see isolines are lines which join points of similar value. Mark the locations of each counting site on the map and record the group results.

Pedestrian flow measurements can be related to distance from the town centre using a scattergraph as shown in Figure 3.26. Usually pedestrian flows decrease away from the PLVI. Alternatively the pedestrian flows can be correlated to shopping quality either using a scattergraph (in this case a direct relationship/positive correlation would be expected) or alternatively a statistical technique such as Spearman's Rank Coefficient.

Shopping quality and street appearance

Many people are attracted to a shopping area by its perceived quality of shops and attractiveness of environment. Table 3.9 shows you how to carry out a shopping quality survey. Shopping quality surveys can be used to examine whether shopping quality deteriorates towards the edge of the core, in the frame or at the CBD fringe. As it is useful to correlate shopping quality with pedestrian densities, it is useful to choose shopping quality survey sites which match

Building height surveys

These can be very useful in large CBDs when trying to delimit the core and the frame. Buildings in town centres are usually taller than elsewhere in order to maximise use of the most expensive land. While you are doing the land use survey, count the number of storeys each building has and record it on your base map. A choropleth technique (colour the highest building the darkest colour) is a useful way to present your finished map.

Pedestrian count

A pedestrian count can be taken to indicate how busy the town is at particular points within the CBD. An appropriate sampling strategy is the key to reliable results. Decide on the positions where counts are to be taken, making sure they are spread **evenly** throughout the town centre, avoiding anomalous sites such as outside a bingo hall at letting out time. The more counts you can do as a group the more reliable your results will be. The time of day during which the counts are taken is a vital consideration; avoid early morning, lunchtimes and late afternoon as many workers and school students will be about.

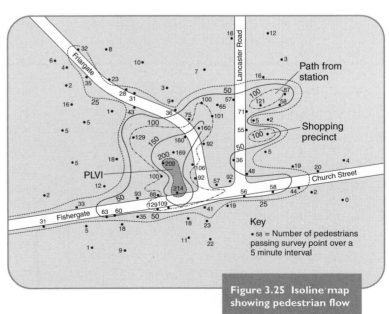

Figure 3.25 Isoline map showing pedestrian flow

Key
• 58 = Number of pedestrians passing survey point over a 5 minute interval

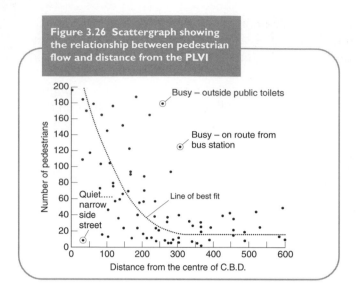

Figure 3.26 Scattergraph showing the relationship between pedestrian flow and distance from the PLVI

the pedestrian density survey sites. It is usual to 'scan' the street at 100 metre intervals along shopping streets using a transect from the PLVI to the edge of the town centre. Street appearance is also useful and can be surveyed in conjunction with shopping quality to achieve a combined value (Table 3.9).

TOP TIPS

Always include a calibration of your quality index with sample photographs in your methodology.

Shopping survey — Table 3.9

Shopping quality		Score (on a scale of 1 to 5)	Street appearance		Score (on a scale of 1 to 5)
A	Type of shop	1 = dominated by department/variety stores, or shops selling 'comparison' goods 5 = wide variety of shop types, convenience goods dominant	E	Safety for pedestrians crossing street	1 = very safe 3 = busy street with pelican crossing 5 = high risk – busy street with no crossing
			F	Shopping crowds	1 = very busy – large numbers of shoppers 5 = very quiet – few shoppers
B	Other land use groups	1 = mainly shops 2 = shops and banks/building societies 4 = mainly offices 5 = very few shops – dominated by houses/industry	G	Street cleanliness	1 = very clean – no litter 5 = very dirty – serious litter problem
			H	Exterior appearance of shops	1 = well-maintained property/attractive window display 5 = poorly-maintained/very drab
C	Retail organisations	1 = national chain stores dominant 3 = mixed – some national and independent 5 = small, independent shop units	I	Traffic/ pedestrian segregation	1 = pedestrianised street/precinct 2 = buses only route 3 = open to all traffic – no parking 4 = open to all traffic – limited parking 5 = main traffic route – no parking restrictions
D	Quality of goods	1 = good quality and/or high price goods 5 = low quality and/or low price goods	J	Vacant premises	1 = all premises occupied 5 = many vacant premises/cleared sites

Shoppers' perception surveys can be used for a number of purposes.

▶ To see where shoppers think the PLVI is.

▶ To investigate shoppers' awareness of particular stores as part of a questionnaire survey.

▶ To see where shoppers think the high quality areas are.

You actually need time to talk to shoppers in the street, so it may be best to try the perception surveys out on family and friends.

Catchment area surveys

These surveys delimit the hinterland of a particular CBD to assess how far people come from to shop or use other services. You can then produce desire line maps (Figure 3.27) to show the catchment area. For one simple question 'where have you come from to shop today?' a grid for recording the answers of 100 shoppers, plus a map to record the place names if you are working in a new area is all you need. Be sure to sample **scientifically** – a random sample of every tenth person usually works well. Some students attempt detailed shoppers' questionnaires to find

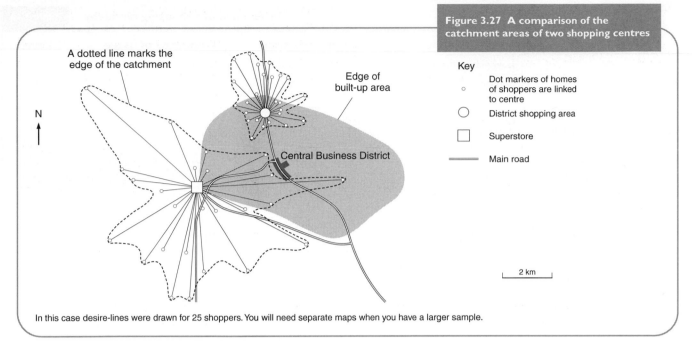

Figure 3.27 A comparison of the catchment areas of two shopping centres

A dotted line marks the edge of the catchment

Edge of built-up area

N

Central Business District

Key

○ Dot markers of homes of shoppers are linked to centre

◯ District shopping area

☐ Superstore

═══ Main road

2 km

In this case desire-lines were drawn for 25 shoppers. You will need separate maps when you have a larger sample.

out what shops people have come to use, how much they plan to spend, or why they come to this particular town to shop. Always pilot the questionnaire and be sure that you really need the results, before you 'pester' shoppers. An alternative method of assessing shop usage is to do pedestrian counts of people entering and exiting from selected shops as a percentage of those passing by. This is known as a **footfall** survey. **Critical path analysis** can be used to track the movements of selected shoppers through the CBD – but make sure you are discreet.

Parking restriction surveys

Parking availability can influence shopping patterns. Usually CBDs are either pedestrianised or buses only, with no parking at all and double yellow lines at the heart of the CBD. Mark the restrictions on your last map devising a classification such as: no parking at any time, parking meters or very restricted parking zone, one hour parking, unrestricted parking, etc.

Shop front surveys

These surveys can be very useful when studying the impact of conservation policies. Look at windows, façades, notices etc. to assess congruence with traditional style.

These surveys are very useful for the **delimitation** of a CBD which forms a very good project for a group.

The impact of change is also a very relevant issue. As you can see from Figure 3.21 most CBDs have zones of **assimilation** where the CBD is growing or

being regenerated with a new precinct, or new specialist shopping area. Equally they have **zones of discard** where decline is taking place, often as a result of out of town developments. These decaying areas are a major problem for planners. There may also be a single change such as the closure of a department store, the impact of pedestrianisation or a new 'edge of town supermarket'. In all cases when you look at change ensure you have old GOAD maps and surveys, or pre-existing projects so you can actually measure changes in shops, shopping quality, pedestrians, and shopping health before and after the new development. Research local newspaper **archives** (usually on 'microfiche' in the local main library) to look at the development of the issues resulting from change. Most town centres/CBDs, in a desire to fight back, employ managers to co-ordinate developments.

SHOPPING STUDIES

Shopping hierarchies

An off-shoot of CBD work is to investigate shopping provision within a town, or district within a very large urban area. In areas serving between 25,000 and 60,000 people there is usually a hierarchy of shopping provision (Table 3.10) based on the order of goods, the size of catchment area and the frequency of usage.

	Order of goods	Catchment area	Frequency of usage	Level
Corner shops	Low – bread, newspapers	Very local – 400 m	Daily	1
Shop cluster of a few corner shops	Similar	Very local – 400 m	Daily	
Neighbourhood shops – a small parade selling mainly convenience goods	Medium – chemists and all basic needs	Local – up to 1 km	Three or four times a week	2
District shops – a major suburban shopping centre	Medium – includes most needs	Up to 3 km	Twice weekly, weekly	3
Central Business District	All, including high order	Very wide – up to 20 km	Weekly, bi-monthly	4
Out-of-town Superstore	All, including high order	Wide	Bi-monthly	
Hypermarket	All, including high order	Very wide – up to 40 km	Monthly	

The features of a shopping hierarchy — Table 3.10

A number of possible investigations include the following.

- Defining the shopping hierarchy within a chosen town.

- Comparing the **shopping health** and **functioning** of one level within the hierarchy.

- Comparing and contrasting the distribution of a series of shopping types or services within a town, e.g. pubs and fish and chip shops.

- Investigating to what extent and why shopping provision in an area has changed over time.

- Investigating the impact of a new superstore on existing shopping provision.

Most local authorities (councils) will have a shopping provision plan and will be able to supply details of where shopping clusters exist. To determine a shopping hierarchy, visit each shopping cluster and carry out comparative surveys to include the following.

- Land use maps

- Pedestrian counts

- **Footfall** surveys

- Critical path analysis

- Shopping quality surveys.

A simple questionnaire (25 people at a neighbourhood area or 50 for a district shopping area is realistic) is a very important part of the survey. Essential questions include.

- Where do you live? How far have you travelled to shop here?

- How did you travel to get here?

- What have you come to buy here? Which shops are you going to use?

- How often do you use this shopping area?

- Shoppers rating based on range of shops, price, convenience, services etc. (devise a 1–5 scale).

You can then refer back to Table 3.10 to make your final decisions based on order of goods, catchment area, frequency of usage and importance.

Distribution of shopping types

Your first decision with this type of study is to decide which shop types you will investigate and how many. You could undertake a corner shop survey (there are numerous local types such as newsagents and fish and chip shops) or you may wish to select contrasting shops as shown in Table 3.11. Use a town street map and local directory (combine Thomson's and Yellow Pages of the telephone directory) to locate them. You will need to analyse the distribution visually and use nearest neighbour analysis to assess whether the shops are randomly, regularly or clustered in their distribution. For common shops you might expect a regular distribution to serve the various neighbourhoods. You could support your survey with **footfall** surveys and basic questionnaires.

A comparison between the distribution of five shop types | Table 3.11

		Town centre	Edge of Central Business District		Edge of inner city areas		Mean distance to town edge	
Betting shops								Mainly concentrated in inner city zones
Furniture								Some concentration in C.B.D. but also in spacious 'out of town' sites
Clothes								Mainly concentrated in C.B.D.
Newsagents								Scattered throughout town, except C.B.D.
Fish and chips								Generally scattered throughout town

Distance in kilometres from the town centre

(Scale: Town centre, 1/4, 1/2, 3/4, 1, 2, 3, 4)

You may find that in some cases the shopkeepers will know (via the paper deliveries etc.) where their customers come from. Some indication of shopping quality is also useful.

INVESTIGATING RESIDENTIAL LAND USE

Getting started

▶ Select an appropriately sized area with clearly defined boundaries, e.g. a post code district, a ward/enumeration district or a particular estate.

▶ A large scale map at 1:2500 or 1:1250 will be needed for your sampling procedure. You sometimes need to sample individual houses and particular streets so an A-Z street finder will also be needed.

There are three very important sources of **secondary data** to help you with residential surveys.

▶ **House price surveys** – These can be obtained via the internet, or directly from estate agents'

advertising literature and can form the basis for a project in itself.

▶ **Council Tax Bands** – There are currently eight such bands, lettered A–H. The main criteria used for banding is the current market value, with band H being the most expensive. You can find out the bands of houses in various streets by going to the Local Council Offices. However council tax bands tend to change quite slowly and rarely reflect extensions and home improvements. This is a problem in recently **gentrified** formerly run down Victorian neighbourhoods where the house value has escalated.

▶ **Websites** – A number of websites designed for house buyers tell you information about each street.

www.upmystreet.co.uk includes detailed assessment of residential desirability.

www.enviro.check includes details of any physical problems such as incidence of flooding, or high levels of radon gas.

These websites are often generalised, but they do conatin some useful background information for quality of life surveys, of house price surveys for your sample street.

TOP TIPS

Develop a sampling strategy which you can justify and which will produce reliable results – for example avoid main road transects for residential surveys as they are not typical.

Basic residential surveys

Age of housing

Figure 3.28 shows a series of photographs showing housing from the following periods.

▶ Pre-Georgian (before 1700)

▶ Georgian 1700 – 1830s

▶ Victorian 1837 – 1901

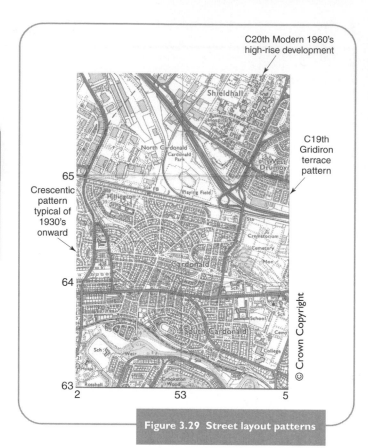

C20th Modern 1960's high-rise development

C19th Gridiron terrace pattern

Crescentic pattern typical of 1930's onward

© Crown Copyright

Figure 3.29 Street layout patterns

Figure 3.28 Types of residential housing

Tudor

Georgian

Victorian

Edwardian

Inter-war

Post-war

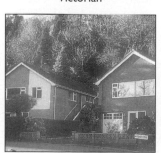
Modern

Very recent

- Edwardian 1902 – 1919

- Inter war 1920 – 1939

- Post war 1946 – 1970

- Modern (post 1970)

- Very recent 1990s

The style of the housing and street layout will help you determine a building's age. Grid iron terraces are characteristically Victorian, whereas crescentic layouts are characteristic of the 1930s or 1950s (Figure 3.29). Some houses may have a date on them. Old maps can also be helpful in dating the age of particular estates as you can find out when they were built.

Building material and housing type surveys

Devise a classification to cover the range of building materials used. The simplest grouping is N = natural stone, M = manufactured (differentiate between brick (b) and concrete/breeze blocks (c)). In some cases a highly distinctive local stone is used and one project idea is to map the distribution of houses using this local stone. It is sometimes useful to include a separate category for roofing (tiles, slates, etc).

Houses can be classified as being:

- terraced

- semi-detached

- detached

- bungalow

- apartments in multi-occupancy buildings

- hi-rise tower block flats

- villas.

Use a system of notation when you map housing types which combines with age and building material (e.g. A5, Ns, Ht Victorian, stone and slate, terraced).

Housing density

This refers to the number of individual houses in a unit area. It can be calculated in the field,

measuring the frontage of each house with a trundle wheel (or pacing) along a defined distance on both sides of the road, e.g.15 houses in 200 m gives a housing density of 7.5 houses per 100 m. Alternatively calculate the density by counting a typical number of units of accommodation in a square kilometre (garden area is included in this calculation). You can also calculate average **garden size** using large scale maps (1:10,000) or bigger.

Housing condition surveys

There are a very useful guide to the state of a neighbourhood. Table 3.12 shows an example of a housing condition survey, often called an Index of Decay. As with any qualitative scale it is subjective, but less so than many such indexes. You have to **sample** individual houses as you cannot assess every house in every street.

You can either sample selected numbers, e.g. 11, 22 and 33 in each street, or use random number tables to select, three houses in each street. Alternatively, you could look down the street and take the worst house, an average house and the best house.

Environmental quality surveys

These surveys can be devised to record the environmental quality of individual streets (Table 3.13). There are many variations on this theme and

Physical condition of buildings, often called an Index of Decay				Table 3.12
No.11 Paradise Street	**None**	**Little**	**Some**	**Much**
Deterioration of walls	0	1	3	5
Paint peeling	0	1	2	3
Displaced roof material	0	1	5	9
Broken glass in windows	0	1	3	7
Broken gutters, etc.	0	1	3	7
Structural damage, e.g. settling cracks	0	3	6	11
Rotting timber	0	2	4	8
Sagging roof	0	2	6	10

Either in the field if time, or on return to the centre

For every street examined, add together the awarded points, *then subtract your total from 60* (**max**)   *60–43=27*

The following general points can be made from your result:

Score	Physical condition of buildings
50–60	Good/excellent
40–49	Satisfactory
30–39	Generally unsatisfactory. May be bad in specific points
20–29	Action needed in near future to improve structure
Below 20	Need to demolish or rebuild

you will need to adapt Table 3.13 to suit your needs, for example to include safety aspects such as street lights.

Fear of crime survey can be usefully done recording the percentage incidence of burglar alarms and neighbourhood watch stickers in each street (use a dot map to present the data).

There are many variations in **questionnaires** you can carry out, usually interviewing people you meet in the street to assess the feel of the community or perceived quality of life etc. The local neighbourhood shopping centre is an ideal place.

Possible investigations within residential areas

Transect studies

Select as many transects as you can from the edge of a town to the edge of the CBD (you need 3–4 people to work on each). In a town with a population of 50,000 – 60,000 people these transects will be about 3–4 kilometres long.

Figure 3.30 shows you how to record the urban transect and to convert it into a map. You will need to do the basic surveys described on page 157, recording age, building material, type and density of housing and garden size. You may choose to sample every fifth house or just record significant changes. Do sample surveys of housing condition and street quality too. It is also worth noting any houses for sale so you can follow this up via estate agents' websites and property magazines in order to get as detailed results as possible for a house price survey.

The transects can be used to test one simple hypothesis (ideal for a basic AS investigation) or alternatively can be put together to contribute to a more extensive A level style of project which looks at the growth of a town over time or builds up a picture of its **urban morphology**. The urban morphology can

An environmental survey						Table 3.23
Feature	**Penalty points**	**Maximum score**	**Feature**	**Penalty points**	**Maximum score**	
Landscape quality Trees and well-kept grassed spaces Few trees and/or unkept grassed spaces No trees or grassed spaces	0 4 8	 8	**Noise** Normal residential standard – quiet Above residential standard – with some noise Main street standard – very noisy	0 2 5	 5	
Derelict land None Small area Large area – a major eyesore	0 4 10	 10	**Air pollution** No offensive smells or obvious air pollution Offensive smells and/or obvious air pollution	0 10	 10	
Litter/vandalism No litter:no vandalism Some litter:or vandalism Very untidy:much vandalism	0 4 8	 8	**Access to public open space** access to park/public open space within 5 mins. (500 m) walk No park/public open space within 5 mins. (500 m) walk	0 3	 3	
Industrial premises All residential properties Some industrial premises Mainly industrial premises	0 5 10	 10	**Access to shops and primary school** Primary school and shops within 5 mins. (500 m) walk Primary school only within 5 mins. (500 m) walk Shops only 5 mins.(500 m) walk No primary school and shops within 5 mins.(500 m) walk	0 2 3 5	 5	
Traffic flow Normal residential traffic Above normal residential traffic Heavy vehicles and through traffic	0 3 6	 6				
			Maximum Penalty Points =		65	

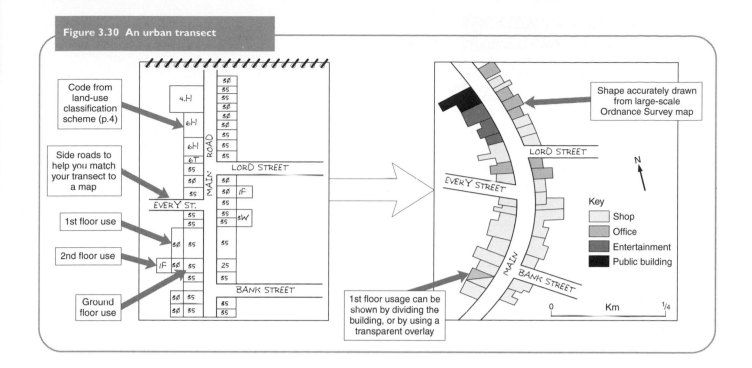

Figure 3.30 An urban transect

Code from land-use classification scheme (p.4)

Side roads to help you match your transect to a map

1st floor use

2nd floor use

Ground floor use

Shape accurately drawn from large-scale Ordnance Survey map

1st floor usage can be shown by dividing the building, or by using a transparent overlay

LORD STREET

EVERY STREET

MAIN

BANK STREET

Key

Shop
Office
Entertainment
Public building

N

0   Km   ¼

then be related to a theoretical urban model such as Manns (Figure 3.33).

Possible hypotheses could include (suitable for single or multiple transects):

▶ The age of housing gets older towards the town centre (Figure 3.31).

▶ The density of housing increases and the garden size decreases towards the town centre.

▶ The quality and condition of housing, and/or that of the environment increases towards the town edge.

▶ House prices increase towards the town edge.

You can combine transects to look at the **growth patterns** of a town. Former census data from 1801 (at 10 year intervals) can be used to draw a graph of **population growth** for the whole town or by individual ward. Figure 3.32 shows you how to relate areal growth with population growth. To find out what factors influence **areal growth** requires an analysis of physical factors such as altitude, slope

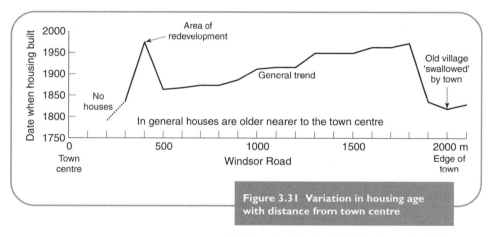

Figure 3.31 Variation in housing age with distance from town centre

and flood liability combined with key economic developments. **Primary** surveys can be done of slope using a clinometer; altitude can be found using large scale OS maps (1:10,000). **Secondary** surveys can be done using www.enviro.check to find out which areas are liable to flooding and subsidence.

Economic factors can also be very important – such as the significance of new communication links (e.g. canals, railways and motorways) or the arrival of a new industry or development such as housing estates for commuters. This requires historical geography research using old guide books and other local history books to make a **time line or chart** of significant events in your chosen town's history.

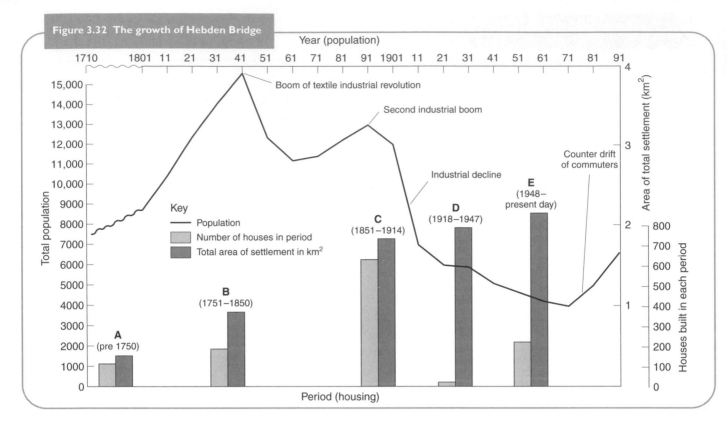

Figure 3.32 The growth of Hebden Bridge

Another use of transects is to build up a map of the age and type of housing, social class and functions found in various parts of a town, and then compare the resulting pattern to an **urban model**, to see how and why it is different or similar. Figure 3.33 summarises the key features of Mann's model which is likely to be useful for most northern English towns. Whilst building age can be established using a standard basic survey, social class of an area is more complex to assess so you need to use secondary census data. The quality of the housing can be assessed by a combination of house type (size), average price, density, garden size, housing facilities such as garages, as well as using average council tax paid for particular districts.

As this is a complex topic it is worth including an evaluation of your method, justifying your methodology and choice of criteria.

Quality of life surveys

These surveys form a very popular topic for investigation. The results are occasionally excellent, frequently very average, and commonly weak. How do you avoid doing a mediocre project?

▶ Define what is meant by quality of life. A useful starting point is to brainstorm with friends a series of factors which you consider constitute a 'good quality of life' (social, economic, environmental etc.) and then rank the factors. There is unlikely to be complete agreement but the following are good indicators or quality of life: access to secure employment, good housing, reliable health services and public transport, well disciplined schools which get good results, green/clean/pollution free environment, lack of crime, peace and quiet, personal space of a garden and garage etc.

▶ Work out your possible surveys to include **primary** and secondary data collection. Primary data could include housing conditions, environmental quality, noise, fear of crime and pollution surveys, open space surveys and access to services, as well as questionnaire surveys on perceived quality of life, and satisfaction with the local district as a

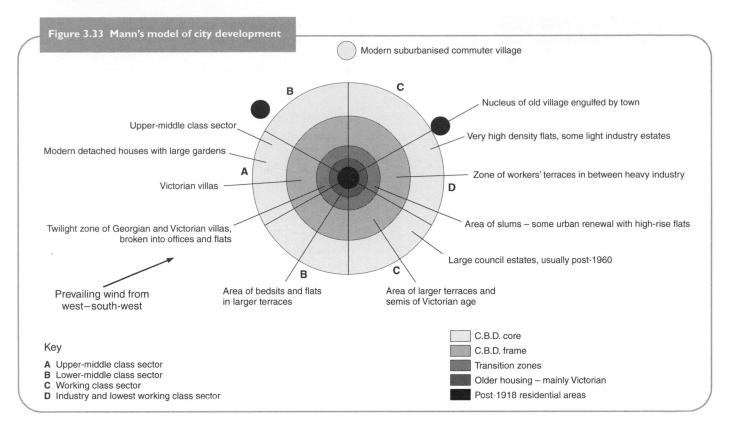

Figure 3.33 Mann's model of city development

Modern suburbanised commuter village

B

C

Nucleus of old village engulfed by town

Upper-middle class sector

Very high density flats, some light industry estates

Modern detached houses with large gardens

A

Zone of workers' terraces in between heavy industry

Victorian villas

D

Area of slums – some urban renewal with high-rise flats

Twilight zone of Georgian and Victorian villas, broken into offices and flats

Large council estates, usually post-1960

Prevailing wind from west–south-west

B

C

Area of bedsits and flats in larger terraces

Area of larger terraces and semis of Victorian age

Key

A Upper-middle class sector
B Lower-middle class sector
C Working class sector
D Industry and lowest working class sector

C.B.D. core
C.B.D. frame
Transition zones
Older housing – mainly Victorian
Post-1918 residential areas

place to live. **Secondary** data could include using a census to develop a quality of life index and to assess social class and ethnicity.

A **social deprivation index** consists of:

▶ unemployment percentage

▶ percentage lone parent families

▶ percentage youth unemployment

▶ percentage households with just a single pensioner

▶ percentage with long-term limiting illness

▶ numbers of dependants (e.g. sick or retired) or children.

A **material deprivation index** consists of:

▶ overcrowding (households with more than 1 person per room)

▶ percentage households with no car

▶ percentage households lacking basic amenities (sole use of fixed bath or WC)

▶ percentage households with no central heating

All obtained from the census.

Additionally people profiles (social class, ethnicity etc) can be obtained from the census.

▶ Decide where you are going to work – anything other than a small area census division such as a ward or an enumeration district poses problems for secondary data collection. Use a census enumeration district map (obtainable from the local library) to mark on the boundaries on a local street map.

▶ Think carefully about a **relevant or contrasting focus** – avoid the obvious situation of comparing the best ward with the worst.

▶ Spend time researching attractive ways of presenting your results. Figure 3.34 shows an interesting way of presenting an environmental quality survey. You could attempt a **composite index** such as that shown in Table 3.14. This is easy to do as it relies solely on ranking. You can also obtain pre-calculated indexes such as Townsend's from the census website

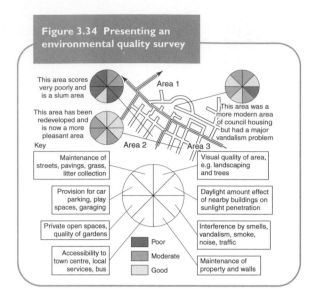

Figure 3.34 Presenting an environmental quality survey

Key
- Maintenance of streets, pavings, grass, litter collection
- Provision for car parking, play spaces, garaging
- Private open spaces, quality of gardens
- Accessibility to town centre, local services, bus
- Visual quality of area, e.g. landscaping and trees
- Daylight amount effect of nearby buildings on sunlight penetration
- Interference by smells, vandalism, smoke, noise, traffic
- Maintenance of property and walls

Poor / Moderate / Good

▶ Where is the zone of greatest deprivation found in X town?

▶ How and why does the quality of life vary in a transect from X to Y?

▶ What impact has regeneration or gentrification made on the quality of life in X area?

▶ Does X really deserve its perceived lowest ranking in terms of quality of life?

▶ To what extent is there a correlation between the index of deprivation and poor environmental quality?

▶ Do areas of lowest quality of life have a poor sense of community?

www.statistics.gov.uk, which you can then relate to your own primary data. Conclude by putting your ward in context, i.e. how does it compare to the average profiles for the town or city you are investigating?

Suggestions for possible projects include the following.

▶ How and why does the quality of life vary in X ward?

House price surveys

These types of surveys can be developed into an excellent project investigation **provided** the title includes scope for some primary investigatory work.

The first stage is to record sufficient house prices. You need an appropriate booking form such as that shown in Table 3.15. You may be able to do most of the work using an estate agent's descriptions, but occasionally you may find an anomalously low or high price – this could be related to domestic

| | | | How to make a composite index of deprivation | | | | | | | **Table 3.14** |

% Ward number	% Households with no car	% Males unemployed	% Households without a bathroom	% One-parent families	% Males employed in manual and unskilled jobs	% Houses rented	% Houses with overcrowding	% Old age pensioners	% Households with ethnic minorities	% Total of ranks
1	6	6	8	8	8	3.5	7	5	8	59.5
2	2	1	1.5	5	1	2	3	1.5	1	18.0
3	1	2	1.5	3	2	1	1.5	3	2	17.0
4	8	8	7	7	6	8	8	8	7	67.0
5	7	7	5	6	5	6	6	4	6	52.0
6	5	5	6	4	4	5	1.5	1.5	4	36.0
7	4	4	4	1	7	1	5	6	5	37.0
8	3	3	3	2	3	3.5	4	7	3	31.5

The information was collected from the small area census statistics.

In all cases a low rank could indicate a bad environment.

Only the number of pensioners proved an unreliable criteria to select.

Property Ref. No.	Location	Distance from town centre	Type	Rateable value	Central heating	Garage	Garden	Notes	Asking price
1	20 Paradise St.	0.4 km	3T	£95	/	/	/	Modernised	£16,000

A house price survey booking sheet — Table 3.15

a perception survey amongst your family and friends as to what factors they think influence house prices. Ask them to rank the following factors in order of importance when buying a house:

- central heating
- garden size
- garage availability
- number of bedrooms (size)
- outside condition of house
- standard of interior decoration
- position of house
- neighbourhood status
- ease of resale
- nearness to basic services.

circumstances, e.g. children wanting to sell an inheritance. It will therefore be necessary to confirm the housing condition by a visit to the house. Figure 3.35 shows you how to present your information and record anomalies.

When recording your data, write a reference number for each house using a 'sticky' coloured spot for each housing type and locate up to 100 houses on each map. If you are comparing wards in a town you will need about five of each housing type for each ward.

A table which adds to your secondary survey by fieldwork, for instance relating the quality of a range of house types to either the actual quality of life (basic survey) or perceived quality of various neighbourhoods (do a ranking survey with a random group of 50 people) is likely to achieve a higher mark. If you stay just with house prices at least begin with

You can correlate your house price survey results with a number of factors such as house type, distance from the town centre, size of garden, council tax band and so on. Use scattergraphs and Spearman's Rank correlation coefficient to analyse your results.

A hot issue at the moment is the impact of a good secondary school on house prices. If you go to one of these state schools with an excellent reputation find out the agreed catchment area of your school according to its admission policy and survey a range of types of houses inside and outside the catchment area. Some parents think it is worth paying up to £15,000 extra to get their children into the 'right' state school without paying fees.

Distribution of recent/new housing developments

Most councils will supply information on housing developments for the last 10 years or so. You need to know both how many houses were built each year and exactly where (a map is useful to locate them). Ideally you need to either survey about five estates or alternatively look at the detailed environmental impact of the building of one new estate.

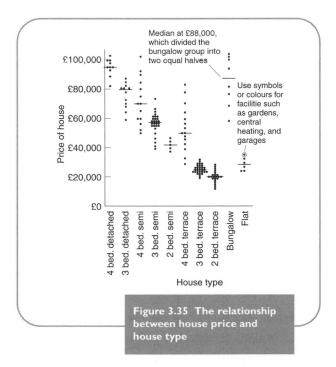

Figure 3.35 The relationship between house price and house type

TOP TIPS

For this title the timing is all important as you need surveys before, during and hopefully after people move in. The process usually takes about a year, so you may need to rely on a previous project for the initial data, or alternatively if your final deadline is approaching and people still have not moved in you will have to 'model' a similar estate.

Survey the type of housing on each estate – size, building material, housing density etc. Record the physical nature of the site in terms of altitude, angle of slope, previous land use etc. Carry out an environmental survey of each estate. Obtain the price range of the houses in each estate either from the estate agents or in the case of very new estates form the building site office. Figure 3.36 shows you how you could present the results.

You could supplement your survey by a simple questionnaire to find out the type of people living on the estate. The basic information needed will include the following.

▶ Where the people moved from.

▶ Why they chose this estate.

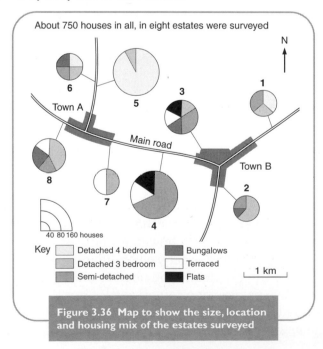

About 750 houses in all, in eight estates were surveyed

Town A

Main road

Town B

40 80 160 houses

Key
- ☐ Detached 4 bedroom
- ☐ Detached 3 bedroom
- ▨ Semi-detached
- ▨ Bungalows
- ☐ Terraced
- ■ Flats

1 km

Figure 3.36 Map to show the size, location and housing mix of the estates surveyed

▶ Brief details on family members, size of family, age, number of children, etc.

▶ Good and bad things about the estate.

▶ Journey to work and type of work.

Keep the questionnaire short and simple. You may be able to get up to the minute secondary data from the 2001 census.

As part of your project, examine the variations between estates, especially public and private housing, or relate house prices to environmental quality.

URBAN CATCHMENT AREA OR HINTERLAND SURVEYS

A town exerts an influence over its surrounding area, by providing employment, shopping, entertainment and services. The extent of this influence is known as the hinterland, sphere of influence, or catchment area.

Delimiting a hinterland accurately and precisely is much harder than you might think, as different services have different hinterlands in terms of size and shape so you have to devise a composite or average hinterland. Also the hinterland of a small town may lie completely within a hinterland of a much larger town for certain services. These are very high order services which require a high **threshold** of customers, from a very long **range**. You can also get distorted results if your sample size is too small, or not representative. Many people you interview have a very limited sense of geography, so having a tracing of a basic road map with you helps to get a more reliable reply.

There are **five** basic types of hinterland survey you can carry out.

1. To ask service providers to delimit their catchment, delivery, or service areas – in some cases this will be a definite administrative boundary, in other cases it is more arbitrary.

2. To interview users of services, for example shoppers, or obtain lists of members or users for example for a high school or swimming club, and to plot their distribution.

3. To develop a specific mechanism, for example for local bus services or the local newspaper to delimit their catchment.

4. If you are working as part of a larger group, you can actually survey people in the villages to find out where they go for their higher order services – for the **Bracey method** (see page 167).

5. To develop **theoretical** hinterlands based on the hierarchy of other neighbouring settlements – for

this you need number of shops and/or population in all the surrounding settlements.

A range of studies are available – a selection is listed below.

---

1. What factors have influenced the size/shape of the hinterland in X town?

2. To what extent do similar settlements have similar sized hinterlands?

3. A comparison of the hinterlands of two smaller settlements.

4. Is X town's hinterland larger or smaller than the one predicted by Reilly's Law?

5. Assess the relative reliability of three methods of hinterland survey.

6. Assess whether contacts between a large urban area and its hinterland decrease as the distance increases.

7. To what extent do shopper surveys provide a fair indication of a tourist hinterland?

8. Has X (a small town of around 2000 people) got a hinterland?

Method	
1	Delivery areas – department store, electrical store, furniture store, local newspaper (to outlets to sell it), postal deliveries.
2	Membership area – sports clubs, social clubs, societies, e.g. Civic Society.
2	Education catchment area – local high school(s), technical college.
1	Employment area – local factory/stress you only need the **general** village/towns in which the workers live, as many firms regard this as confidential data.
1	Service area – **public services** – police, ambulance, fire station, library/mobile library, postal sorting office (postcodes), employment office, hospital.
	private services – telephone exchange, water, electricity (the Yellow Pages may help here as they sometimes show maps).
1	Administration area – parliamentary constituency, local council district, tourist information area.
2	Shoppers catchment – interview up to 200 people – random sample outside shops and services, asking people in which settlement they live.
3	Local newspaper – number of methods – record the settlements mentioned in news articles, e.g. WI reports village news and sports team news. Small advertisements and for sale columns can also be used **but** there are snags. Plot each settlement mentioned, but be aware of rogue results.
3	Local bus services – obtain a timetable to record the settlements which are directly linked to the town by a bus service, the journey times and the number of bus journeys on a standard day. Note with dealing with rural bus services many settlements may not have a service.

TOP TIPS

Choose the right size of town – a market town is ideal with at least 5000 people and a significant level of service provision.

Choose the right mix of higher order services. A range of about 15 services should work well as it represents the variety of reasons people come to a town. Low order services, e.g. food shops and primary schools, should not be selected. A high school and a supermarket are much more likely to have a significant catchment area.

Writing up

Produce a series of maps comparing the sizes and shapes of all the hinterlands studied. Discount any obviously anomalous ones. Even so, you will notice a great range of size and shape. Size is partially related to the order of the service, but shape may be influenced by a number of factors such as other larger settlements, access or physical barriers as hill or a border.

It is important that you try to define an overall sphere of influence which can be done in two ways.

1. Composite hinterland (Figure 3.37) In this case the boundary is drawn visually where the most lines overlap.

2. Average hinterland (Figure 3.38) The distance is measured from the settlement to the furthest boundary of each hinterland at the eight main compass points.

The size and shape of a hinterland can be calculated **theoretically** using **break point theory** according to Reilly's Law (Figure 3.39). Select the appropriate settlements from a map (the nearest rival centres in all directions). The census data will give you population information, and county council data should give up-to-date figures for the number of shops.

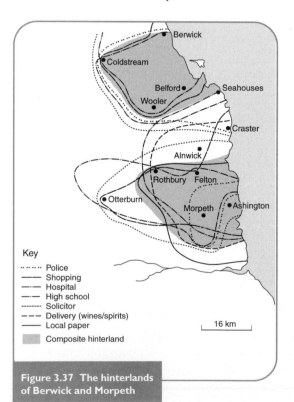

Key

.... Police
— — Shopping
—·— Hospital
—·· High school
······ Solicitor
— — Delivery (wines/spirits)
——— Local paper
░ Composite hinterland

16 km

Figure 3.37 The hinterlands of Berwick and Morpeth

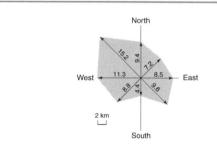

To delimit the 'average' hinterland								
Type of hinterland	Distance from Newtown to hinterland boundary (km)							
	N	NE	E	SE	S	SW	W	NW
Veterinary practice	20	7	10	9	6	6	13	7
Bank	8	6	6	9	1	1	2	15
Solicitor	8	7	8	9	3	5	10	10
Travel agent	9	6	6	5	5	7	9	10
High school	6	4	4	3	3	4	5	10
Delivery area – furniture store	8	16	25	24	5	21	33	22
Circulation area – local newspaper	8	6	6	5	5	8	14	18
Police	9	6	6	5	5	5	8	24
Postal area	9	7	6	3	5	5	9	19
Swimming pool	9	7	8	24	6	26	10	17
Total distance	94	72	85	96	44	88	113	152
Average distance	9.4	7.2	8.5	9.6	4.4	8.8	11.3	15.2

Figure 3.38 Delimiting the average hinterland of Newtown

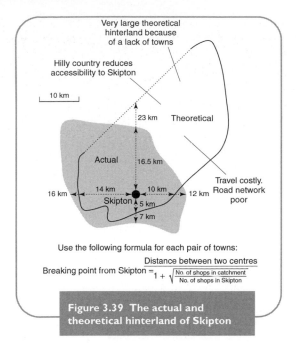

Very large theoretical hinterland because of a lack of towns

Hilly country reduces accessibility to Skipton

10 km

23 km   Theoretical

Actual   16.5 km

16 km   14 km   10 km   12 km

Skipton   5 km

7 km

Travel costly. Road network poor

Use the following formula for each pair of towns:

$$\text{Breaking point from Skipton} = \frac{\text{Distance between two centres}}{1 + \sqrt{\dfrac{\text{No. of shops in catchment}}{\text{No. of shops in Skipton}}}}$$

Figure 3.39 The actual and theoretical hinterland of Skipton

Next measure the distances between the chosen settlement and all the selected settlements, and apply the breaking point formula.

The theoretical hinterland can then be drawn in and compared to the results from fieldwork. Figure 3.39 shows a worked example for Skipton, North Yorkshire. Note that whilst the west, south and eastern boundaries are broadly similar, suggesting there is competition from towns such as Keighley and Clitheroe, in the north the theoretical area is much larger than in reality, because of the apparent absence of rural centres. The

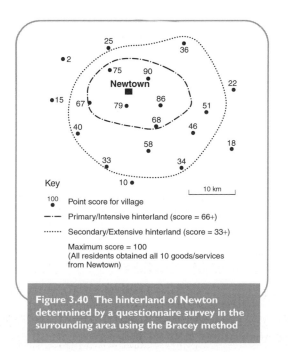

25

36

2

75   90

Newtown   22

15   67   86

79   51

68

40   46

58   18

33   34

10

Key

10 km

100 • Point score for village

—·— Primary/Intensive hinterland (score = 66+)

········ Secondary/Extensive hinterland (score = 33+)

Maximum score = 100
(All residents obtained all 10 goods/services from Newtown)

Figure 3.40 The hinterland of Newton determined by a questionnaire survey in the surrounding area using the Bracey method

narrow winding roads, hilly country and generally sparse population, mean that few trips are made to Skipton, with the easier trip to Richmond, Ripon, or the 'upmarket' Harrogate being preferred.

An alternative approach, the **Bracey Method**, is to visit a range of villages to questionnaire a sample of residents as to which town they use for higher order services. Whilst it is possible to get a definitive limitation of both primary and secondary hinterland (Figure 3.40) there are a number of issues to overcome. Logistics can be a nightmare – check public transport availability beforehand, or recruit a co-worker with a car as you may need to survey 20–30 villages. Sampling residents can also become an issue – many villages are very quiet during the daytime, with only a skewed sample of female pensioners on the street.

You will need to interview a **random** sample of 10 people asking them where they usually go for 10 goods and services. In theory the maximum points score for each place is 100. All settlements with 2/3 (66/100) of the maximum score are placed in the primary hinterland, and those with at least 1/3 (33/100) of the maximum score) are placed in the secondary hinterland. The lower the score the less that village relies on the town you are investigating, and the more services it obtains from elsewhere.

It should be possible to graphically (scattergram) or statistically correlate (Spearman's Rank coefficient) the percentage of people shopping in the town under investigation with distance from that town. You may find that the **distance-decay** of the town's influence does not decrease at a uniform rate as distance increases, but a clear inverse relationship is likely to occur.

POPULATION AND PEOPLE SURVEYS

As shown in Figure 3.41 there are a number of very interesting areas of research you can undertake involving people and populations. As over 80 per cent of the population of Britain live in urban environments, the ideas are included in this section; however, many could be applicable to a rural environment.

The main source of data for these enquiries is the **census**, in particular the data for wards and the smaller enumeration districts. Using census data is **secondary** research. As many exam boards require you to collect primary data in the field it is 'safer' to

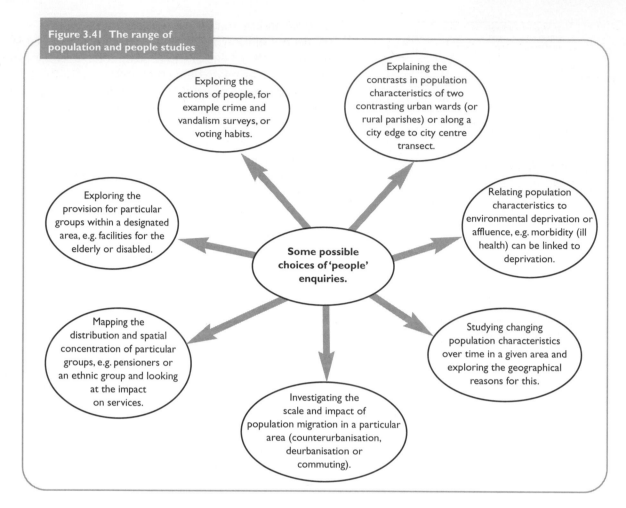

Figure 3.41 The range of population and people studies

Explaining the contrasts in population characteristics of two contrasting urban wards (or rural parishes) or along a city edge to city centre transect.

Exploring the actions of people, for example crime and vandalism surveys, or voting habits.

Relating population characteristics to environmental deprivation or affluence, e.g. morbidity (ill health) can be linked to deprivation.

Exploring the provision for particular groups within a designated area, e.g. facilities for the elderly or disabled.

Some possible choices of 'people' enquiries.

Studying changing population characteristics over time in a given area and exploring the geographical reasons for this.

Mapping the distribution and spatial concentration of particular groups, e.g. pensioners or an ethnic group and looking at the impact on services.

Investigating the scale and impact of population migration in a particular area (counterurbanisation, deurbanisation or commuting).

link up your census analysis with some primary research, for example assessing environmental deprivation (see page 162).

From the 2001 census you can find basics such as the following:

▶ Population totals (density can be calculated).

▶ Age and gender composition (to draw pyramids).

▶ Marital status and family size (important to identify single parents etc.).

▶ Ethnic proportions and religious allegiance.

▶ House type and tenure (with details of facilities, e.g. central heating).

▶ Density per room in houses.

▶ Standards of education.

▶ Employment and therefore social class.

▶ Degrees of morbidity (self declaration of ill health).

▶ Car ownership.

Such information can be used to develop an index of socio-economic conditions of both the ward and the people living in it.

Also available at ward level is a deprivation index available for every ward and local authority in England for the year 2000. You can get access to a number of indicators such as health, disability, education, geographical access to services and housing as well as a **single deprivation score** for each on www.statistics.gov.uk.

Examples of studies

Contrasting wards
The basic data for a study involving contrasting wards can be obtained from secondary census data as

indicated above, but a more in depth analysis can be achieved by visiting your chosen wards to carry out surveys such as housing condition or environmental quality (see page 157). Ideally you need to develop a hypothesis which links one or more recorded census characteristics with the situation on the ground.

For example, your hypothesies could be 'Poor quality environments are lived in by disadvantaged groups in X ward'. In such a study you would first need to split the ward into enumeration districts. You should then identify the disadvantaged groups (e.g. pensioners, families with no adult employed, single parent families with very young children). Finally you will need to relate the figures to the results of housing condition and environmental quality surveys.

Alternatively, you could take one piece of information from the census, for example 'Levels of morbidity increase in areas with a poor physical environment' or 'Levels of educational attainment vary with the quality of the environment'.

For the first hypothesis you need to collect percentage morbidity figures from a number of enumeration districts (say around 10–12). These are usually available from your local library. Data may also be available from your local Health Authority or Community Health Group, especially as is the case with many deprived areas where a Health Action Zone (HAZ) has been developed. **Mortality statistics** are also available by local government area from the Registry of Births, Deaths and Marriages. Figure 3.42 shows the range of interesting cartographical techniques you might use to portray your results.

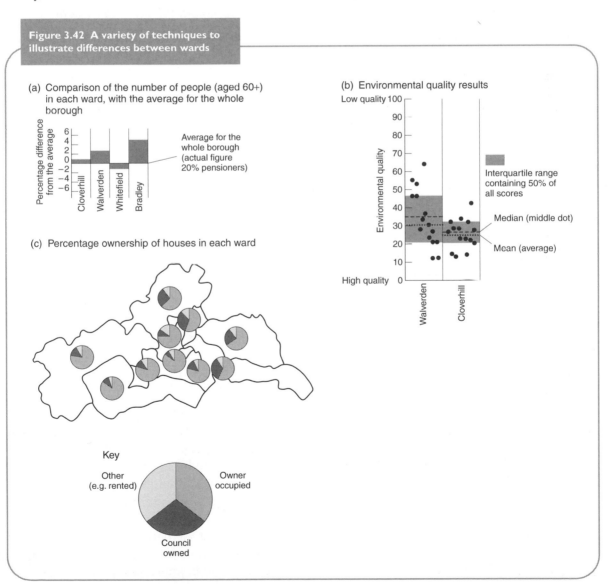

Figure 3.42 A variety of techniques to illustrate differences between wards

(a) Comparison of the number of people (aged 60+) in each ward, with the average for the whole borough

Percentage difference from the average

Average for the whole borough (actual figure 20% pensioners)

Cloverhill Walverden Whitefield Bradley

(b) Environmental quality results

Low quality 100

Environmental quality

High quality 0

Interquartile range containing 50% of all scores

Median (middle dot)

Mean (average)

Walverden Cloverhill

(c) Percentage ownership of houses in each ward

Key

Other (e.g. rented)   Owner occupied

Council owned

For educational attainment you may just be able to work with your school centre to correlate GCSE scores to postcodes and then link this to environmental quality and a number of census characteristics.

Changing population structure/characteristics over time

The first stage in a population study is to decide on a clearly defined area, for example an inner and outer ward, for which a full set of small area census statistics are available for the period from 1801 (the first census). Some areas will be more suitable than others – a fact which can be seen from a pilot survey. Choose areas such as those shown in Figure 3.43, which show marked contrasts, and for which there have been minimal boundary changes. You can also contrast **population pyramids** for different time periods. You may find features resulting from changing demographics and

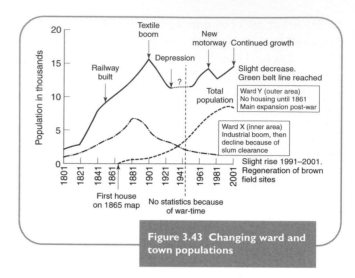

Figure 3.43 Changing ward and town populations

changing migration periods, such as reduced birth rates or out-migration for employment.

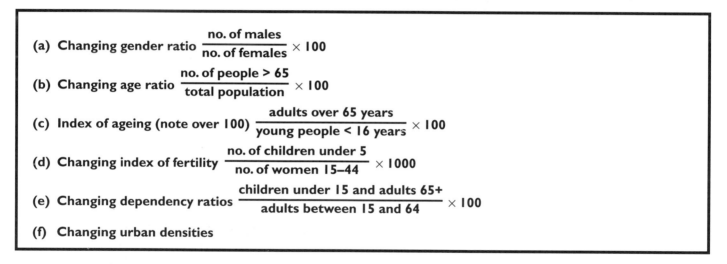

(a) **Changing gender ratio** $\dfrac{\text{no. of males}}{\text{no. of females}} \times 100$

(b) **Changing age ratio** $\dfrac{\text{no. of people} > 65}{\text{total population}} \times 100$

(c) **Index of ageing (note over 100)** $\dfrac{\text{adults over 65 years}}{\text{young people} < 16 \text{ years}} \times 100$

(d) **Changing index of fertility** $\dfrac{\text{no. of children under 5}}{\text{no. of women 15–44}} \times 1000$

(e) **Changing dependency ratios** $\dfrac{\text{children under 15 and adults 65+}}{\text{adults between 15 and 64}} \times 100$

(f) **Changing urban densities**

In addition to enumerators records for any historic census surveys, visit your local county record office to ask for details of

▶ **Parish registers** which provide a historical analysis of a population. Baptism registers give details of births and usually the occupation and addresses of the parents. Marriage registers can show population movement as they record the home parish and occupations of both the husband and wife. Burial registers provide dates, ages and causes of death.

▶ The 1986 Doomsday survey is available from local libraries – it provides a mine of information, but be **selective**.

▶ Electoral registers are available for every year and give addresses of every declared voter and can be used to show any nation. Note only a certain percentage of people agree to allow their names to be on the published registers.

To go with your secondary research you must also undertake primary fieldwork such as a housing survey of the age and nature of the house types (some may be vacant or multi-let in inner city areas or there might be a major building programme in a particular ward). This can be combined with questionnaire surveys on randomly sampled households. Find out the reasons for the particular changes of population structure discovered, but only use questionnaires in neighbourhoods where you are well known.

One interesting, but restricted investigation focus (it has only occurred in certain areas) is **gentrification**. This is the process by which working class inner city areas change over to a middle class occupancy with a notable improvement in environmental quality. For such a project you will need to obtain enumeration district census data so that you can map social change since the 1960s. This will show any changes in socio-economic groups, car ownership, housing amenities, housing tenure and density. You should then conduct a housing condition and environmental quality survey in the districts which you have identified as being the most gentrified. Note any evidence of modernisation (e.g. new roofs, window frames, doors and extensions). Carry out a **services** analysis recording obvious middle class services such as wine bars, delicatessens and book shops. If you have a series of historic services surveys before gentrification at your centre these are an invaluable source of secondary data. Use a questionnaire to interview a sample of people to see where people have come from and how they are employed, in gentrified and non-improved houses. Review house prices in the area, and compare with an unimproved area, also try to compare house prices with at least one set of historic records from five years ago.

Migration studies

Migration can be investigated from a long term point of view, i.e. permanent migration, or a short term daily basis, for example commuting to work or school.

Whilst general information exists in **migration tables**, usually for large towns over 50,000, to get precise patterns of movement requires questionnaire surveys of about 50 sample households. For long term migration you will need details of the following.

▶ Age/gender of all family members.

▶ Type and place of work.

▶ Place of birth of all family members.

▶ How long the family has lived in the house.

▶ Where the family came from originally and why.

▶ What advantages/disadvantages there are in living there.

▶ Names and destinations of family members who have left the area.

Figure 3.44 shows how you can use a ray diagram to show local movement (known as a kinship survey) if only family members are shown. Figure 3.45 shows how you can use a histogram to show more widespread movement.

For short term day to day migration it is worth asking people where they go to for a number of higher order services and any visits they make to family and friends over a period of one month. You could compare the average movements of people by age bands, or by social class or by village. Mark the position of each completed questionnaire with a dot on a map of your chosen area, to check you have a representative sample of all housing types. Ideas for projects involving daily movement include the following.

▶ School area catchment survey.

▶ Journey to work.

▶ Journeys to recreation and leisure.

The first example of a school area catchment survey is described in detail below as it is such a

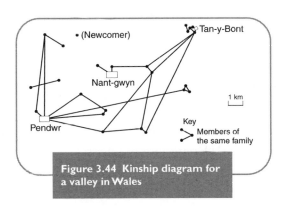

Figure 3.44 Kinship diagram for a valley in Wales

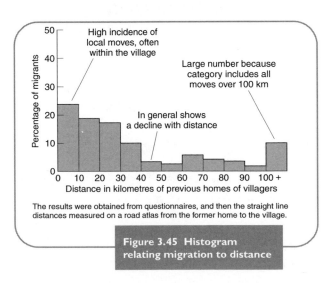

Figure 3.45 Histogram relating migration to distance

viable project for those who may be prevented from doing outside fieldwork.

School area catchments can be very interesting, as parents and children have the right to choose a suitable (secondary) school for their children. In many urban areas where a lot of schools in a small area are competing for pupils, explaining the resultant patterns of movement makes an interesting study.

Stage 1 Carry out a perception survey with parents to determine what makes a good school. It may be related to one or more of the following factors.

▶ Results (league table position).

▶ Quality of teaching (inspection report).

▶ Standard of discipline.

▶ Facilities for sport/music.

▶ Quality of care (e.g. anti-bullying measures).

▶ General reputation.

▶ Nearness to home. A school with a high reputation may have a much larger catchment.

Stage 2 Get permission from your own school to get the home addresses (just streets) of all the pupils. A fall back would be to get **postcodes only** if the registers are considered confidential. It is possible to use a computer program which can locate postcodes within enumeration districts so you can get an insight into the socio-economic characteristics of the pupil's home area.

You can either look at issues caused by the journey to school, such as car parking surveys, school bus surveys, traffic noise, pupil walking routes and litter, etc. or you can compare the catchment area of your school with a neighbouring school. Formal permission will need to be sought from the headteacher, explaining who you are and why you want the information. Alternatively it may be possible to look at the geographical reasons for a changing school catchment area (change of status, redrawing of boundaries, rising reputation etc.). Table 3.16 shows an example of a suitable booking form for this.

To present your results first record your official school catchment area, using an acetate or tracing paper overlay on a map and marking the location of each student's home as precisely as possible – use different colours for each year group. You can use dot maps, flow maps, or methods such as those shown in Figure 3.46.

| School transport survey | | Table 3.16 |

Home address give street or village _____

Year group _____

	Coming to school	Returning home
Main method of transport		
Distance travelled		
Time of departure		
Time of arrival		
Total journey time		

Note: can be made more sophisticated, e.g. by adding journey times

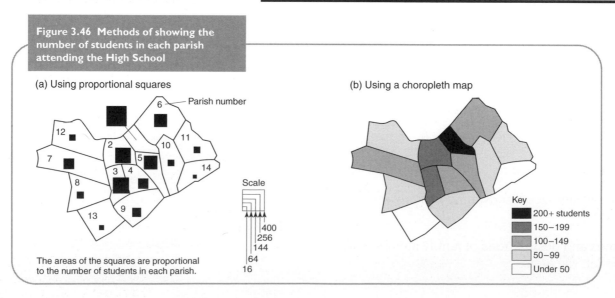

Figure 3.46 Methods of showing the number of students in each parish attending the High School

(a) Using proportional squares

Parish number

Scale
400
256
144
64
16

The areas of the squares are proportional to the number of students in each parish.

(b) Using a choropleth map

Key
200+ students
150–199
100–149
50–99
Under 50

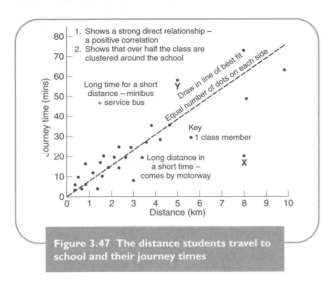

You can then analyse journey patterns, looking at distance travelled, modes of transport used and relating journey distance to time, as shown in Figure 3.47.

Analysing the distribution and degree of spatial concentration of designated groups

In a multi-cultural society there are many ethnic minority groups. The 2001 census actually asked people to identify which ethnic group they belonged to, so such information will be available down to enumeration district level. Limited information is also available historically, for example old electoral registers. Ethnic communities are frequently clustered, and often initially formed as part of a major migration. West Lancashire, for example, contains numerous people of Irish descent, who came to seek work in the nineteenth century after the Potato Famines. You can correlate the concentration of particular ethnic groups to particular housing types, or locations such as inner city areas, which provide access to a range of employment. In some cases ethnic groups are forced to concentrate in deprived neighbourhoods with poor environmental quality.

At a street scale **electoral registers** can be extremely useful for the identification of certain groups. Within some religious groups such as Sikhs, Hindus and Muslims, names follow set rules and are easily recognisable. The Commission for Racial Equality publishes a very useful guide to recognising names. For some communities such as Afro-Caribbean the technique will not work as names are often indistinguishable from other groups. Polish and Irish names are also relatively easy to recognise. You can then work out the **segregation ratio** (SR) for each street.

$$SR = \frac{S \text{ Number of CEG (chosen ethnic group) occupied buildings}}{\text{Total number of occupied buildings}}$$

Figure 3.48 shows the results of an SR survey for part of Bradford. Note a ratio of 1:0 would represent a street whose houses were all occupied by one CEG and 0:0 would represent a street whose houses were all occupied by no CEG. **Nearest neighbour statistics** can be used to provide a measure of the clustering of any ethnic community. In areas with a high concentration of a particular CEG you could survey the availability of specialist facilities such as banks, travel agencies, specialist shops, religious buildings and entertainment centres to assess the extent to which the group are provided for.

You may decide to visit the 20 streets/areas with the highest segregation ratios, to record the age, type, condition, price of houses and environmental quality. It is best to get co-operation from community leaders and Race Relations Officers before conducting your surveys to ensure you handle your enquiry in a sensitive way. Compare your results with 20 streets which contain no or very low concentrations of ethnic minorities. Minority groups are frequently initially

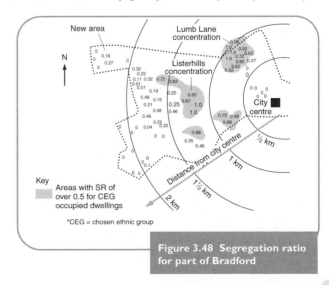

Figure 3.48 Segregation ratio for part of Bradford

Figure 3.49 Distribution of pensioners

Enumeration District (ED)	Percentage of pensioners in each ED	Percentage of total population in each ED	Location Quotient (LQ)
1	40	25	(40 ÷ 25) = 1.6
2	30	30	(30 ÷ 30) = 1
3	25	20	(25 ÷ 20) = 1.25
4	5	25	(5 ÷ 25) = 0.2

A Location Quotient above 1.0 shows above average concentrations.

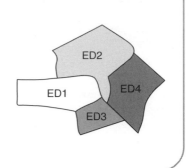

forced to live in the poorest quality housing and the most unfavourable conditions. Using historic data it is possible to track how particular groups have moved over time often because of **filtering**.

Figure 3.49 shows how the Location Quotient (LQ) is used to indicate the degree of concentration of pensioners in various districts. You can analyse the nature of the neighbourhoods in which high concentrations of pensioners are found and try to explain why this is the case. You can also relate this distribution to specialist pensioner services such as nursing homes, and mobility advice centres.

Exploring the provision of services for a particular group of people such as the disabled, homeless, travellers or elderly people.

In order to begin such a study make contact with your local council who will almost certainly have a designated officer in charge of your particular group who you can discuss the major issues with.

Possible studies on this subject include the following.

▶ How user friendly is the town centre for a person in a wheelchair, or blind people?

▶ How well provided is a settlement for pensioner services?

You will need to talk to representatives from your chosen group, for example walking along a parade of shops with a blind person to record all the hazards. **Investigating** access for the disabled involves carrying out an audit based on the following criteria.

▶ arrival and parking

▶ entrances to buildings

▶ entering facilities

▶ circulation inside.

Take a walk around your local shopping centre and using the checklist below undertake an audit of:

▶ How many disabled people there are in relation to non-disabled people.

▶ How people with physical and sensory impairments would get to the shopping centre (e.g. is public transport accessible, are there disabled parking spaces?).

▶ How accessible the areas between the shops are for people with physical and sensory impairments (e.g. are there steps but no ramp, is there lots of street furniture?). You can do a mobility survey to assess the suitability of the sheet for wheelchair access, looking at surface, gradients, kerbs, steps, pavement width, congestion, etc.

▶ How accessible, both from the street and once inside, the shops are to people with physical and sensory impairments (e.g. are the aisles wide enough for a wheelchair, are there lifts between floors?).

▶ How the shops provide for people with learning and development disabilities (e.g. are the visual signs easy to understand, are items on sale colour coded?).

▶ How many specialised services are provided in the immediate area (e.g. is there a shop mobility scheme, are there accessible toilets?).

The checklist below should help you focus on specific areas. You can grade each on a scale 0 (no facility) to 5 (excellent provision).

	Score		Score
Arrival and parking ▶ Well signposted and easy-to-find car parking for disabled users. ▶ Designated car spaces for disabled people that are close to the CBD. ▶ Trained staff available to help disabled people (with signs to indicate so). ▶ Accessible path from car park to building (e.g. dropped kerbs). ▶ User-friendly path for people with sensory impairments (e.g. tactile paving). ▶ Obstacles (e.g. bollards/street furniture) highlighted by colour contrast and tactile surfaces.		**Entering facilities** ▶ Appropriate height of reception desk. ▶ Adequate seating. ▶ Publicly accessible disabled toilets. ▶ Map of site including levels of accessibility. **Circulation areas** (for shops, public areas): ▶ Adequate directional signage (tactile as well as visual). ▶ Corridors wide enough. ▶ Level fire exits. ▶ Suitable floor surface. ▶ Tactile paths/guides.	
Entrances to building ▶ Provision of both steps and ramp. ▶ Hand-rails provided on both sides of steps/ramp. ▶ Doorbell can be reached by all. ▶ Audible/tactile/visible intercom. ▶ Easy-opening door. ▶ Level threshold across doorway. ▶ Door width sufficient to allow wheelchair access.		**Circulation inside** ▶ Lift large enough to accommodate wheelchair. ▶ Doors open wide enough. ▶ Appropriate height of control panel. ▶ Appropriate alarm/phone height. ▶ Audible and visible signage. ▶ Suitable dimensions of treads/insets of stairs. ▶ Hand-rails to both sides of stairs and in contrasting colours.	

Figure 3.50 Accessibility map

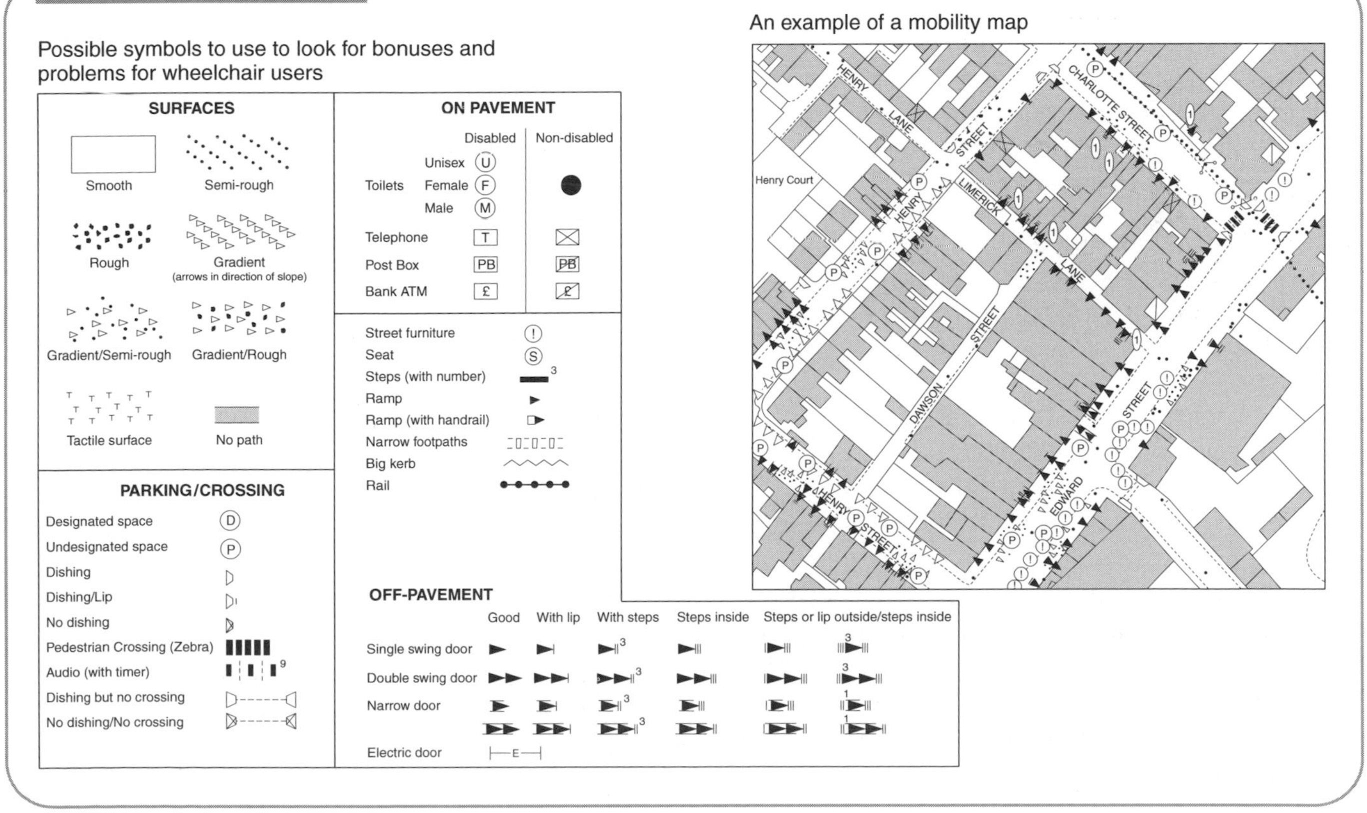

Possible symbols to use to look for bonuses and problems for wheelchair users

An example of a mobility map

You can draw a number of maps to show your results (e.g. Figure 3.50 shows a detailed map of accessibility made as a result of a mobility survey).

Getting disabled people to draw their own perceptual maps of their local town centre is also very useful as you can correlate their perceptions of trouble spots with your findings (see Fig 1.4 page 16).

Table 3.17 lists some very useful resources for carrying out disability surveys.

Investigations on the geography of crime

There are a number of possible studies on this issue, many of the investigations are closely inter-linked.

▶ Assessing susceptibility of houses to burglary.

▶ Analysing crime trends: where, who, when.

▶ Fear of crime. Perception surveys of where and when.

▶ Investigating vandalism or the incidence of graffiti.

▶ **Susceptibility to burglary** Design characteristics and the location of certain premises may encourage crime. Equally, design improvements to buildings, e.g. gated communities or the impacts of security measures such as security lighting, burglar alarms, and CCTV cameras may lead to a decrease in crime. Possible investigations include the following.

▶ Is the incidence of crime related to the design and location of houses/town centre shops and services?

▶ Is the incidence of crime related to the level of security measures – again at house or town centre level?

You will need to devise an index of burglarability as shown in Table 3.18.

Presentation can be in the form of maps to show the distribution of burglar alarms etc. and the results of burglarability index surveys. Aim to show the relationship between the number of crimes committed

Table 3.17
Disability Unit in local town hall. Lists of facilities, details of all provision in shopping.
Shop mobility for local town.
Disability Research Unit.
Disability Net – www.disabilitynet.co.uk
Centre for Accessible Environments – www.cae.org.uk
Ability Net – www.abilitynet.org.uk

and the design, location of premises and degree of security provision.

Analysing the geography of crime The first stage in this study is to obtain crime data, all of which will be for **reported** crime as opposed to **actual** crime. **Nationally** the British Crime Survey, or the Criminal Statistics for England and Wales will provide such data. Crime rates are recorded for a county or metropolitan district, or police area as a rate per 100,000 population to facilitate comparisons. **Regionally** the local police station should yield some detailed statistics by town, or even village area. The Community Liaison Police Officer, or the Crime Prevention Officer are good first lines of enquiry. Table 3.19 shows a typical set of statistics from a Neighbourhood Watch Association.

Locally it is difficult to get any form of statistics other than by scanning the local paper for crime watch, court reports or crime story reports, and this has to be done over a period of 1–5 years depending on the levels of incidence of crime.

Index of burglarability		Table 3.18
	House	Retail/commercial premises
Provision of burglar alarm	10	10
External lighting system	5	10
Security cameras		10
Security doors with shutters		10
Security windows with shutters	5	10
Metal bars across windows	5	10
Other security design features (e.g. blocking off alley ways)		10
Security system at entrance – guard/dogs		10
Neighbourhood watch sticker	5	
Backing on to open space	−5	
Quiet street	−5	
Secluded entrances – bushes etc.	−5	

Local crime statistics	Table 3.19

Legend: **A** – Burglary, Dwelling; **B** – Burglary, Other;
C – Criminal Damage; **D** – Theft from M/Vehicle;
E – Theft of Pedal Cycle; **F** – Theft of M/Vehicle

	A	B	C	D	E	F
Adderley		1		1		
Astley	1	6				
Baschurch	1	9		2		1
Cheswardine		1		1		
Child's Ercall				1		
Clive		1				
Cockshutt		2				
Ellesmere Urban	9	12	1	10	6	8
Ellesmere Rural	3	4		4	2	
Grinsill		2			1	
Hadnal		3		1		
Hinstock	3	3		1		1
Hodnet	2	6		3		2
Hordley						
Ightfield		3				1
Loppington		2		3		1
Market Drayton	44	50	1	84	16	22
Moreton Corbet		2				1
Moreton Saye		1		1		
Myddle		8		1		1
Norton in Hale	3	1				1
Pim Hill	1	10		13		1
Prees	2	9		10		4
Shawbury	1	2	1	2	1	2
Stanton on Hine		2				
Stoke on Tern	1	8		1	1	1
Sutton on Tern		2		2		
Welshampton & Lyneal	2	2		3		
Wem Urban	8	21		12	8	12
Wem Rural	5	4		2		4
Weston under redcastle				6		
Whitchurch Urban	12	53	3	22	12	29
Whitchurch Rural	4	10		9	1	6
Whixall	2	2		2	2	3
Woore		2		2		2

There are a number of theories which you can test out about crime distribution and criminals.

▶ Is crime concentrated in the CBD, or in rich or poor residential areas? (Relate to house prices etc.)

▶ Are crimes more common in secluded areas or busy areas? (Relate to house and street design using large scale maps and distance from a main road.)

▶ Is more crime committed at night (see timing of offences) or in the winter (darkness factor)?

▶ What is the impact of security measures and house design (relate to index of burglarability) or Neighbourhood Watch Schemes?

▶ Are criminals clustered in the worst neighbourhoods? (Social exclusion factors of decaying inner city terraces or badly kept council estates.) Visit the 10 streets where most criminals live and carry out housing condition and environmental quality surveys – compare the results to 10 streets which are completely free of convicted criminals.)

▶ Do criminals live in the wards which appear to be most socially deprived?

▶ Is there a relationship between age, gender, nationality and criminal activity? (You can relate the distribution of criminals to population structure and composition.)

TOP TIPS

The key to success of this type of project is to map and classify the crimes accurately. The initial logging from the newspapers is tedious but the results are very interesting.

The fear of crime The way to undertake a project on this theme is by perception surveys – either of town centres or residential areas. Ask various people to identify safe and unsafe areas on your maps (use a stratified sampling technique) and then use a Likert scale to assess people's attitudes to selected safe or unsafe areas identified from your survey and the reasons for their views. A Likert (attitude) scale involves providing a respondent with a series of statements and asking them to respond on a four or five point scale.

For example:

I would feel very safe alone here in the dark.
Strongly agree ☐   Agree ☐
Disagree ☐   Strongly disagree ☐

You can then start looking at a correlation between security devices such as CCTV and lights to assess their importance, and the 'zones of fear' and safety can be correlated with the actual incidence of crime.

Vandalism The main problem with a study of vandalism is to devise a way of evaluating the severity of each occurrence of vandalism. The following can be taken into consideration:

▶ writing on walls (graffiti)

▶ smashing of lamps and windows

▶ dumping of hostile litter, e.g. tin cans and broken bottles

▶ the breaking of fixtures such as seats

▶ damage to trees/plants.

You will need to choose two neighbourhoods and carry out initial surveys to record, classify and locate various instances. Highly vandalised areas are often quite hostile places so never work alone.

When you have devised your own scale (an example is given in Table 3.20) do not forget it is a very **subjective** scale and you should calibrate it with photographs. Having done this, use the small area census statistics to relate the incidence of vandalism to deprivation and other factors such as

unemployment. In theory the most disadvantaged people will have the most need to show their boredom and frustration. You can also relate vandalism to housing quality (price, density, condition and environmental quality), accessibility and visibility of site.

Investigating voting patterns

This can be an interesting study especially for those students interested in politics. Election results vary considerably both by parliamentary constituency at a general election, and locally at council elections both spatially and over time. Whilst the geography of voting patterns at a national scale is very interesting, for an A level project local election data are the most appropriate because you can support your enquiry with primary research too.

First decide the scale of your investigation, e.g. the whole town, or for a large city a transect of wards across it. Obtain the voting results for **each ward** (the voting unit for local elections) from the public library, the **Town Hall** or the local offices of one of the main political parties. As a vote is a secret ballot you may only be able to get area data. Classify the results by ward in terms of **voting preference, swing** from the last election and **turn-out** (per cent voting). A useful technique is to use **proportional pies** for each ward so you can show the precise details of the voting preference.

Unless you are working in an area with only one or two wards you will need to **sample** from the wards to carry out further primary fieldwork on the socio-economic and environmental characteristics. You could investigate strong preferences for **any** party, or marginal areas which have changed, or alternatively reasons for variations in turnout. A number of stereotypes exist about voters, for example traditionally the Labour Party is thought to have gained the bulk of its votes from working classes, younger people, people living in council housing, ethnic minorities and from disadvantaged groups such as the unemployed. The traditional Tory voter is perceived of as being middle class, managerial or self employed, owning a house in a wealthier district, and possibly more likely to be female and older. You can test these traditional stereotypes by using a combination of small area census data, housing price and condition, and environmental quality surveys in the most **polarised** wards. Alternatively you could link your enquiry to lowest turnout and the factors which lead to this.

Graffiti assessment chart		Table 3.20
		Score
Maximum size of words	0 – 10 cm	+1
	11 – 25 cm	+ 2
	26 – 50 cm	+ 3
	> 50 cm	+ 4
Maximum size of pictures	0 – 10 cm	+ 1
	11 – 25 cm	+ 2
	26 – 50 cm	+ 3
	> 50 cm	+ 4
Nature	Obscene/racist	+ 5
Method	Ink/pencil	+ 1
	Felt pen	+ 3
	Aerosol	+ 5
	Wood carving	+ 8
	Concrete drilling	+ 10
Visible from	1 – 5 m	+ 1
	6 – 10 m	+ 2
	11 – 30 m	+ 3
	31 – 50 m	+ 4
	> 50 m	+ 6

TOP TIPS

Notice that a questionnaire approach has been avoided because voting is a personal issue the results are likely to be disappointing. However it is worthwhile writing official letters requesting interviews with agents, party officials, or councillors in order to gain background information on local voting patterns and any current issues which may have affected the vote (e.g. the threat to close a local hospital or raise council tax by 10 per cent). If you are analysing change over time, be sure to look at national trends too, as they can sometimes complicate local election results, particularly if voting was on the same day for both elections.

WEBSITES

Office for National Statistics, for all demographic data: www.ons.gov.uk

Ordnance Survey and Multimap for downloadable maps: www.ordsvy.gov.uk
www.multimap.co.uk

Good maps: www.experian.com/intl/uk/office

POLLUTION OF THE ENVIRONMENT

a)

b)

c)

Pollution and its impact on the environment is a very popular choice of fieldwork investigation at A level. Although you can undertake a general pollution survey, for example studying the sources and possible impacts of pollution in a small mining village, it is better to focus on one aspect of pollution such as atmospheric, water, noise or land pollution. Water pollution has been dealt with on page x, the other three aspects are covered in this chapter.

At first sight pollution surveys seem to rely heavily on sophisticated and expensive equipment.

However, well chosen titles, combined with a range of straightforward fieldwork investigations can lead to some very successful and interesting projects. Secondary data are usually readily available from environmental monitoring stations. For **legal** reasons you are strongly advised to avoid a project title which assesses the impact of a single point pollution source such as a named factory chimney, or overflow. To carry out this type of fieldwork you would need the full permission and co-operation of the factory or power station.

INVESTIGATING AIR POLLUTION

Table 3.21 shows the main constituents of air pollution, their common sources, as well as their impact on health and the environment. You might want to look at one type of pollution such as that from solids, or link your survey to particular weather conditions or the type of relief and land use in your chosen study area. The scale of the air pollution problem varies tremendously at different times of the year and in certain places. For example, it is often very bad during calm weather, in the bottom of valleys or beside very busy main roads. You can focus on a particular cause such as the impact of traffic, or explore the impact on people's health of differing pollution levels. People's perceptions of the causes and levels of pollution also add an interesting dimension to a study. Some suggestions for project titles are outlined on page 182.

Primary data collection

In this section you will find details of how to carry out basic pollution surveys. One of the key decisions to make is how to select an appropriate **sample**.

▶ If you are looking at a **single point** source, for example a quarry or a thermal power station, you will need to locate your individual sampling points at set distances (e.g. 250 m) away from the source in **all** directions. Draw a simple compass diagram with eight cardinal points on your map to locate possible sampling points.

▶ In the case of a **linear pollution** source, such as traffic flows along a main road, you will need to set up measuring stations at fixed distances along the road, and also at fixed distances **away** from the road.

▶ Where you are assessing **general pollution** levels and how they vary within a given area, you will

The main constituents of air pollution | Table 3.21

	Pollution	Main sources	Environmental impact	Health impact	Primary data investigation	Availability of secondary data
SOLIDS	Dust particles	Ash, soot Diesel	▶ Low level ozone formation ▶ Blackening of buildings	Impact on lungs and throat	Leaf washing Dust traps Dust strips	Particulates Volatile Organic Compounds Old photographs and maps of sources
	Smoke	Chimneys	Poor visibility	Impact on lungs and throat		
	Lead	Leaded petrol	Poisonous	Accumulation in body can cause brain damage	Ringlemann Chart	
GASEOUS	Sulphur dioxide and nitrogen dioxide	Fossil fuels	Winter smog Acid deposition	▶ Constricts air passages ▶ Infiltrates bronchial tubes ▶ Asthma	Lichenography Acid Rain Survey using narrow range pH strip Passive monitoring using diffusion tubes (Drager) CO meters (available from Bedfont Scientific M39 7HN) Car speed surveys to assess traffic impact for CO CO_2 NO_x	SO_2 levels NO_x levels
	Carbon monoxide	Exhaust fumes	Poisons vegetation	Interferes with oxygen supply to heart and brain		CO] levels CO_x
	Carbon dioxide	▶ Fuel combustion ▶ Fossil fuels ▶ Power stations	Enhanced greenhouse effect			
	Ozone	Road transport	Forms smog	Chest, nose and throat problems	Ozone detection cards (Dryden EH10 3JQ) Watch experiments	Ozone O_3

need to use a grid system of sampling. You can use random numbers to generate a sample site in each square or systematically sample grid intersections. The size of the grid square will depend on the size of the urban area and whether you are working individually or with a class group.

▶ For a very **large** urban area, or for investigating contrasts in relief and land use a **transect** across

the area or from the rural urban fringe to the city centre. Always be sure to record the **precise** location of each sampling point (give a grid reference). It is also very important to collect full environmental details of each survey site, for example relief, aspect and altitude, as well as land use and any possible pollution sources. Details of weather – wind speed and direction, rain and temperature – should also be recorded, as well as the time of day.

Project titles	Tips for success
1. How and **why** air quality varies in a transect from X to Y.	Ideally join up with your friends to work together on a number of a transects.
2. How and why air quality varies in x small town.	Group work will be needed for a **large** town as otherwise the coverage is difficult.
3. How does the nature and level of air pollution vary between a rural area and an urban area?	You will need to select two grid areas or two contrasting wards – be guided by the availability of secondary data.
4. What impact does the weather have on air pollution (type and levels) in an area?	A longer term investigation, perhaps taking three sample days in a year. You could compare a depression sequence with a summer high.
5. An investigation which considers whether valley areas are more polluted than upland areas in a town and why.	You will need to select an area with varied relief.
6. To what extent does air quality agree with people's perception of the problem in a city/town?	This requires extensive questionnaire work.
7. An assessment of the impact of a stretch of urban road/motorway on pollution levels.	This requires **detailed** and accurate measurements. Check you have adequate equipment.
8. What impact do traffic flows have on air quality?	This combines pollution measurements with modelling of traffic impact.
9. What impact does a high level of pollution have on morbidity (ill health) and mortality (death)?	Choose an area with highly polluted regions.

▶ It is often very useful to support your pollution survey with a **questionnaire** about local people's perception of pollution. Figure 3.52 shows you a possible framework which you can adapt for your specific purposes. You will need to organise your sampling procedure so that your questionnaire results can be correlated with your survey of pollution levels. You need to keep the **length** of the questionnaire to a minimum to ensure a good 'take up'.

Solid particles

Particles of ash, soot, dust and smoke result from combustion, usually from industrial chimneys or, before the days of Clean Air Acts and smokeless fuel, from coal fires. Another source is from diesel, especially from heavy lorries. Diesel has been promoted as an environmentally friendly fuel, but that does not take into account these solid particles known as **particulates**.

The most reliable method of **dust** sample collection (only available when deciduous trees are in leaf) is to collect a sample of five similar sized leaves at head height from a tree near each of your sampling points. Wash the dust from the leaves at each site onto filter paper and visually record the amount (use a qualitative scale **or** arrange the filter papers in rank order of dust coverage for statistical analysis). Where significant dust pollution is recorded, allow the dust sample to dry on the filter paper, weigh it, and use a microscope to measure the particle diameter size. You may then be able to work out the nature of the material.

It is also possible to develop **dust traps**, which are necessary if you wish to study the impact of wind direction on air pollution. All methods have some

Figure 3.52 Pollution questionnaire

AIR QUALITY STUDY QUESTIONNAIRE

Good Morning: I am a student from Wedgewood Sixth Form College researching the quality of air in the area. I would be grateful if you would answer a few questions for me. It will only take a few minutes.

1. Do you live in this area?   YES ☐   NO ☐

2. How would you rate the dust levels in this area?

    1 ———————→ 10
    extremely bad/high levels   very good/low levels

3. How would you rate the scale of fumes, e.g. petrol?

    1 ———————→ 10
    high levels   low levels

4. How much is smoke a problem in this area? Is it
    (a) Great problem   (b) Some problem   (c) Not much of a problem
    (d) No problem   (e) Don't know.

5. Have you noticed if problems of air quality are worse in certain types of weather? e.g. in the winter or if the wind changes?

    If so, please specify.

6. Are there any respiratory problems in your family?   YES ☐   NO ☐

7. To what extent do you believe the air quality in this area can affect such problems?
    (a) A great deal   (b) Some   (c) Not much
    (d) None at all   (e) Don't know.

8. Do you think the air quality in this area represents a risk to people's health?
    (a) A great risk   (b) Some risk   (c) Little risk
    (d) No risk   (e) Don't know.

9. Do you think air quality has improved over the years?
    (a) Yes, a great deal (b) A little   (c) Not much
    (d) Not at all   (e) Don't know.

10. Do you think the problem could be reduced in any way? If so, how?

11. For how long have you lived in this area? _____ years approximately.

12. Male/Female

13. Estimate Age   15/21
                    22/50
                    50 +

problems, not least finding dry, vandal-proof sites. One of the more effective methods is to make traps to cover the **four** sides of a cube (12 cm²) placed in a **fixed** location. To make a dust trap mount a strip of wide sellotape at least 10 cm in length with the sticky side upwards on white card, pin one of your dust traps on to each face of a cube so that north, west, south and east faces can be compared. Unless the area is very heavily polluted you will need to keep the strips in place for a week. Make sure you record the wind conditions (speed and direction) each day. Display the strips according to aspect so as to compare the amounts of dust collected visually.

Smoke

Emissions of smoke can be classified using a **Ringelmann** chart (Figure 3.53). Use the chart under daylight conditions once you have mounted it on a piece of white card. Hold the chart in a vertical plane in a line between yourself and the smoke coming out of a chimney.

Compare the darkness of the chimney smoke with the scale on the chart and record the Ringelmann number which is the best match. You can use intermediate estimates if the smoke colour is between two Ringelmann numbers. Record the time of day, size of smoke plume and weather conditions. You may decide to do a series of measurements to relate smoke formation to weather conditions.

Sulphur dioxide

Sulphur dioxide (SO_2) is one of the most serious air pollutants and is emitted principally as a result of fossil fuel usage.

The most realistic method of SO_2 assessment is to use **lichenography**. Lichens are simple plants which are found on trees, walls and other surfaces. As lichens have no roots they absorb moisture and solutes through their entire body surface and are very sensitive to the effect of SO_2 pollution. In very badly polluted environments no lichens may survive, but in general it is possible to use a pollution scale related to the presence of a particular lichen. Different lichen species tolerate different levels of pollution, and therefore certain species act as useful indicators of levels of SO_2 pollution (Figure 3.54). You should only record the highest zone reading for any spot, because the cleaner the air the more indicators of **all** types are likely to be found.

Sampling lichens has to be done rigorously to get reliable results. Lichens have very specific ecological requirements or **niches**. For example, their distribution is affected by aspect, pH, surface material and climate, so all these factors must be considered when locating study sites. Lichens are very slow growing, so avoid new buildings – look for old walls and graveyards. When sampling lichens you will need to look at similar heights and aspects, as well as using an acetate grid to measure percentage coverage of the zone indicator lichen. You can use a **choropleth** map to display lichen

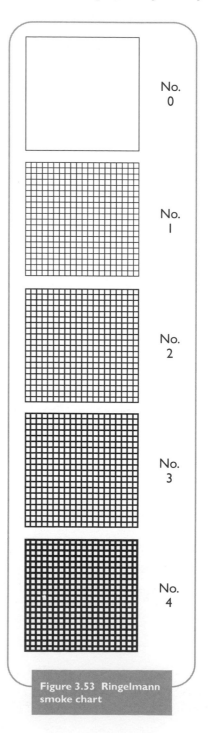

Figure 3.53 Ringelmann smoke chart

zones at survey points using dark shading for where there are no lichens.

Nitrogen dioxide

Nitrogen dioxide (NO_2) can be measured by passive monitoring using diffusion tubes. These can be purchased from local science laboratory equipment suppliers – your chemistry department will almost certainly know of a source of supply. The local Environmental Health Department may be able to help.

Ideally you will need to site your tubes at a variety of places in your study area to get a detailed coverage. They need to be mounted (at about 2 m above the ground) on buildings and lamp posts. If you are looking at the impact of traffic on pollution you will need to find sites at measured distances from the main roads with quieter residential areas to provide a contrast. A **monitoring** period of a week is probably the most suitable, but you should also carry out a second set of measurements, for example at a different time of the year or during different weather conditions. At the end of the monitoring phase you can send your **carefully labelled** tubes to your local contact analysis services where, for a small fee, the results will be processed.

You can present your results as isoline maps or located bar charts.

Acid rain

SO_2 and NO_2 emissions are the main causes of acid rain. You can do a very useful **acid rain survey**, which **indirectly** indicates levels of pollution. Collect rainwater in a rain gauge (see page x for how to make your own simple one) and test for acidity with either a pH meter or litmus strips. You **must** clean out the rain gauge each day with distilled water to avoid contamination from previous readings as rain water is always slightly acidic. You also need to avoid contamination from leaves, especially pine needles, so the siting of your rain gauges is important. An acid rain survey is useful for comparing three or four locations such as a rural setting, an outer suburb, an inner city area and an industrial zone.

Ozone

At ground level ozone is harmful to plants and humans. It is the main constituent of smog, and is produced when nitrogen oxide (from car exhausts)

Figure 3.54 A lichen chart used for pollution

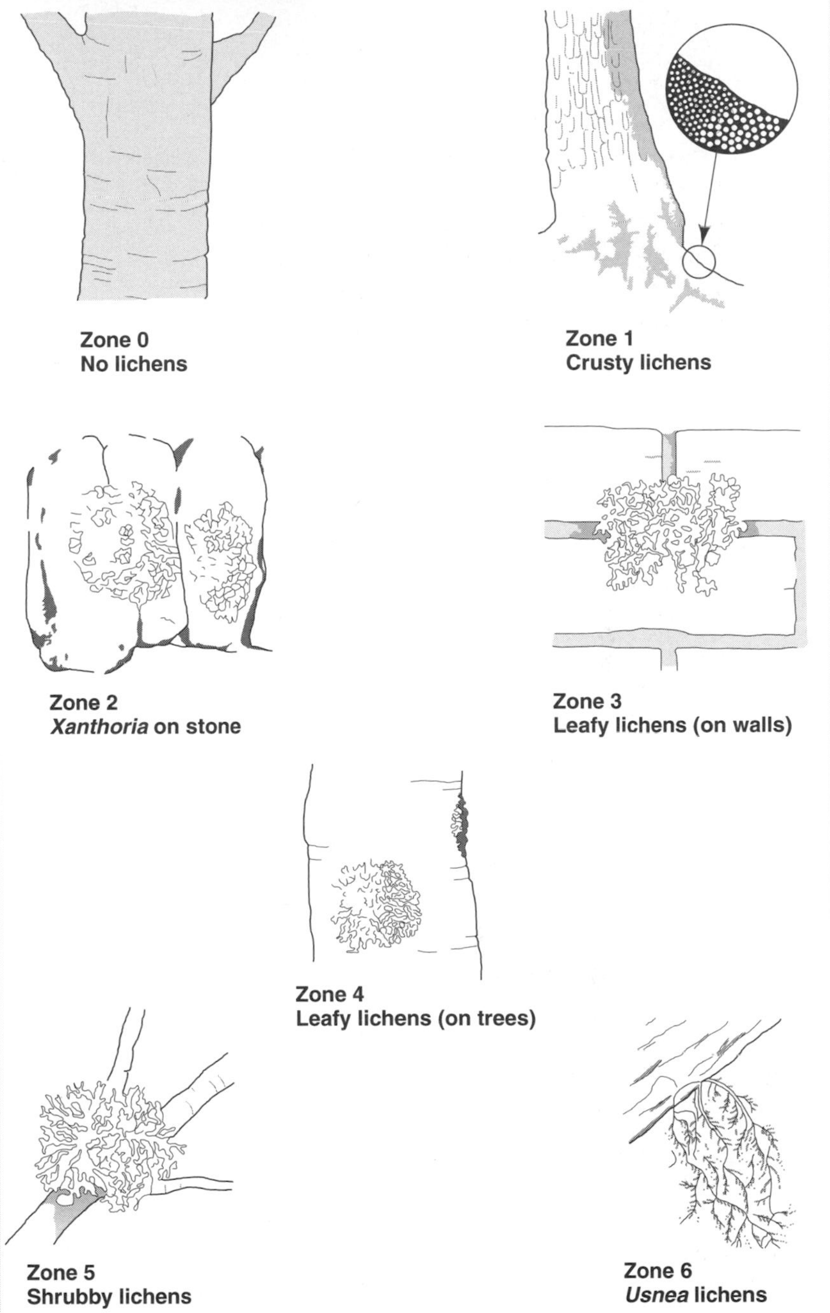

Zone 0
No lichens

Zone 1
Crusty lichens

Zone 2
Xanthoria on stone

Zone 3
Leafy lichens (on walls)

Zone 4
Leafy lichens (on trees)

Zone 5
Shrubby lichens

Zone 6
Usnea lichens

Lichens are biological indicators of sulphur dioxide pollution. Zone 1 indicates the most SO_2 pollution, whilst zone 6 has the least pollution. Note: lichens are found in different locations and will be influenced by climate, moisture, aspect etc, in addition to air pollution. (Full lichen chart from the Field Studies Council)

reacts with oxygen in the presence of sunlight. You can conduct a study on the assessment of plant damage due to ozone. Ozone enters leaves through the stomata and attacks the cells causing brown spots. Obtain some seeds and plant them in a variety

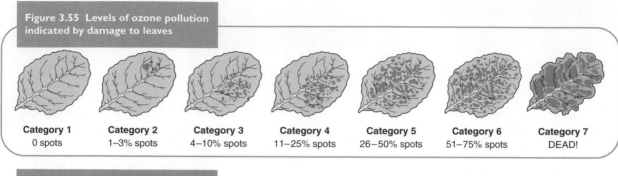

Figure 3.55 Levels of ozone pollution indicated by damage to leaves

Category 1	Category 2	Category 3	Category 4	Category 5	Category 6	Category 7
0 spots	1–3% spots	4–10% spots	11–25% spots	26–50% spots	51–75% spots	DEAD!

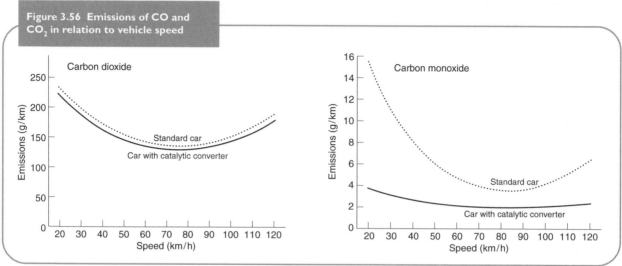

Figure 3.56 Emissions of CO and CO_2 in relation to vehicle speed

of locations. Once the plants have grown you can use a visual scale (Figure 3.55) to assess the level of damage to them by looking at the size and density of the brown spots. Both higher levels of ozone and longer periods of exposure will lead to more damage and ultimately the death of the plant.

A simple way of measuring air pollution

If you are doing an urban pollution project and are looking at how traffic density and speed can contribute to the production of gaseous pollutants and general air quality, a **car speed survey** can provide some useful indications of levels of carbon monoxide (CO) and carbon dioxide (CO_2) (Figure 3.56). You will need to record the speed in kilometres per hour of a sample of ten different vehicles covering a distance of 100 m within a given time period in the vicinity of each of your pollution measuring sites. Use the following formula:

$$\text{Speed (V)} = \frac{3600 \times 100}{\text{time taken to cover 100 m}}$$

$$\longrightarrow \text{then } \frac{V}{1000} \text{ to convert to kph}$$

Once you have found an average speed, you can then use Figure 3.56 to work out the likely emission levels. Surveys at rush hour will show slower moving traffic and this can cause poorer air quality. It is useful to record the type of traffic (car/lorry) and whether they use diesel or petrol as well as the overall flow.

Secondary data

Local sources of secondary pollution data include the following.

1. **Environmental Health Department** This department has responsibility for environmental monitoring. You may be fortunate enough to have a local monitoring station which will record most of the following (nitrogen oxide, nitrogen dioxide, sulphur dioxide, carbon monoxide, ozone, particulates, volatile organic compounds and lead). Recordings are made daily, often in conjunction with basic weather data. The data can be related to synoptic chart weather information, and also used to look at seasonal variations in pollution levels. You can also contrast weekday

levels with those on Sundays and national holidays. **Historic** data may also be available so you can compare current pollution levels with the past, for example in the 1950s before the passing of the Clean Air Acts.

If you are relating pollution levels to **health** you may also be able to obtain data on the incidence of heart, lung and respiratory disease statistics by ward. You can then correlate this information with your surveys and questionnaire work.

2. **Environment Agency** This organisation is responsible for regulating waste disposal, releases to the air from some industrial processes and safeguarding the water environment. They can provide maps showing estimated annual mean concentrations of NO_2, SO_2 and particulates, as well as some data on above average ozone concentrations for your local area. Information is usually collected by river catchment area or by urban area.

TOP TIPS

Investigate what sources of secondary data are available **before** you choose the precise title of your project as secondary data support is very important for most investigations.

INVESTIGATING NOISE POLLUTION

Noise pollution **by itself** is not a common choice for an investigation. Usually it is just one part of a wider environmental impact or nuisance survey. However, if you have access to a reliable and accurate noise meter (decibel recorder) noise pollution levels could form the basis of a very focused AS project.

Some possible projects on this topic include the following.

▶ How do noise levels decrease away from a motorway/main road?

TOP TIPS

The key to success is a rigorous and effective sampling strategy so you can collect a good coverage of readings to make an **isoline** map of noise levels. In particular you should portray the effects of **distance decay**.

(Compare a busy stretch of road with a normal road.)

▶ How do noise levels vary around an airport?

▶ An assessment of the success of noise management techniques at an open cast pit (e.g. concrete barriers, trees, earth mounds).

▶ The impact of a problematic development (e.g. beach disco, large pub complex) on a neighbourhood.

▶ How and why do noise levels vary within a village?

You will need to make a note of weather conditions – in particular wind direction and speed. People who live by a noisy feature will tell you what a difference wind speed and direction make.

If you do not have a decibel recorder you can use the qualitative scale shown in Table 3.22. However this scale is very subjective and therefore best suited to noise surveys as part of a wider environmental impact survey, e.g. of a football ground, leisure complex, or quarry.

Qualitative noise scale — **Table 3.22**

Score	Decibels (db)	Description
7	140	Deafeningly loud, e.g. Concorde/jumbo taking off
6	120	Very loud, e.g. a live pop band
5	100	Loud, e.g. a heavy lorry passing by
4	80	Noisy, e.g. main traffic routes at busy times
3	60	Quiet, e.g. a normal conversation
2	40	Very quiet, e.g. a library
1	20	Barely audible, e.g. a ticking watch

Possible title	Visual impact	Air pollution survey	Dust	Noise	Smell	Traffic flow survey	Litter survey	Impact on ecosystems	Subsidence survey	Water pollution	Micro-climate survey	Thermal pollution	Impact on drainage	Perception survey
Impact of an open cast mine	•	•	•	•		•	•	•					•	•
Impact of an incinerator	•	•	•		•	•	•	•						•
Impact of a deep mine	•	•		•		•			•	•			•	•
Impact of a landfill site	•			•	•	•	•	•						•
Impact of a thermal power station	•			•		•					•	•		
Impact of an industrial plant, e.g. cement works	•			•	•	•		•					•	•
Impact of a quarry	•	•	•	•		•		•		•			•	•
Impact of a wind farm	•			•		•		•						

Support your practical data collection with questionnaire surveys, of people's perception of noise levels (use stratified sampling at various distances away from the noise).

LAND BASED POLLUTION

There are numerous possible investigations involving land based pollution, some examples are given below.

Most investigations look at a range of land based pollution as part of an overall environmental impact survey, but there are a number of opportunities which focus on a particular aspect (e.g. litter or derelict land issues). Others such as **smell** which relies on a subjective 'sniff' scale (0) = no smell detectable (1) = barely noticeable – just detectable rising to (5) = highly offensive leading to a very strong reaction (almost keeling over, pinching nose, holding breath). Visual quality is also much better as part of a larger survey.

Litter surveys

Litter is a major issue in many locations within Britain. Some possible ideas for investigations which concentrate **only** on littler are given below.

▶ How and why do litter levels vary in a community?

▶ Where are the main problem sites for litter in a community?
  – could be linked to a fast food issue or a school or even a wedding issue (outside churches & Registry Offices).
  – What impact do litter bins have on litter levels? There is a school of thought that claims that litter bins (especially over/flowing ones) actually generate litter.

▶ Is fly-tipping a major issue in the area? (Survey old quarry sites, streams etc.)

▶ What impact has the opening of a complex had on litter levels? (You will need to record litter levels before and after the complex opened – see Figure 3.57).

When undertaking a litter investigation you will need to record the following information.

▶ The number of pieces of litter per unit area. This will enable you to measure the litter density – this can be shown using a choropleth map (Figure 3.58).

▶ The volume in terms of weight (use a spring balance in the field).

Figure 3.57 Litter levels before and after a restaurant and cinema complex had opened

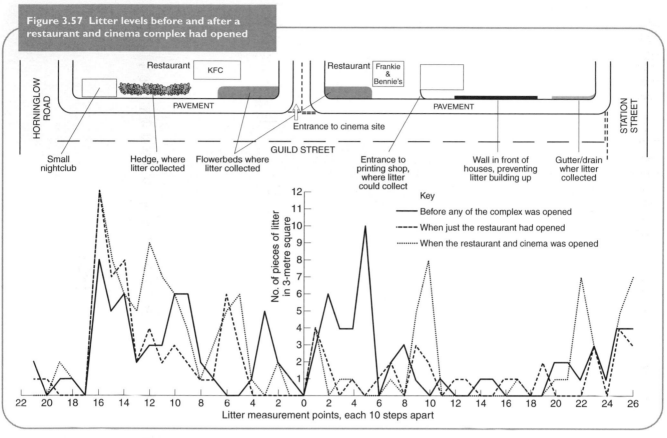

Figure 3.58 Choropleth map showing litter density

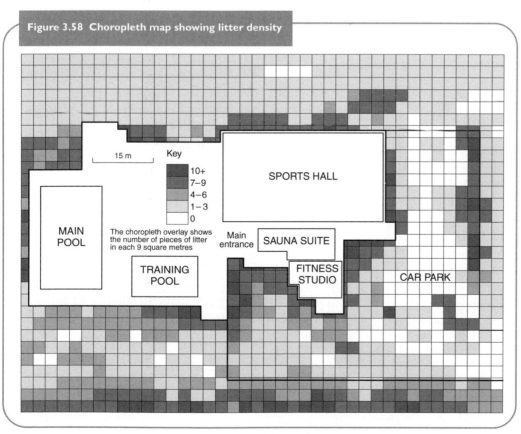

▶ The content of the litter – estimate the percentage of paper, glass, silver foil, rubber, tin cans etc. for each measuring site. You can use proportional charts to show your results.

You can support your surveys with proposed management strategies and questionnaires on perceived amounts of litter and possible causes.

For many projects, litter surveys will just form part of a survey, for example when considering the impact of a market or a honey pot tourist site.

Figure 3.59 Vegetation in a derelict colliery site. Photos can also be used for Visual Impact Surveys

Derelict land and buildings surveys

The importance of using derelict land, i.e. brownfield sites, has become a national issue. Some derelict sites are very unpleasant, but others have been transformed into ecological havens. A number of very interesting titles are available as outlined below.

▶ What would be the best use for a specified derelict site (e.g. an old quarry or the site of an old school)? Ideally you will need the Council plans which will indicate a range of possible uses. You can then research the likely impacts of such uses.

▶ An assessment of the potential of **six** derelict sites in an area. You should record the sites on a large scale map and investigate the potential use of each site. You will need to measure and rank your sites in terms of size, shape, access to site, any negative factors such as poisoned ground, general readiness of site, e.g. services available, ownership/cost of land to buy, adjacent land use, etc. Alternatively you could look at a specific proposal such as a new supermarket, and make an assessment of a number of possible sites.

▶ Assessment of the ecological significance of a derelict site. Figure 3.59 shows a derelict area of a colliery site (itself now closed too) with a wide range of vegetation which has colonised the site. You should use quadrat to measure vegetation height, cover, distribution of species, species count and calculate an index of diversity.

When investigating a derelict site, you will need to undertake surveys on the landscape quality (including litter surveys) and also record people's views on the value of the site.

Table 3.23 shows an example of a scale of visual pollution which is particularly appropriate to use in an assessment of derelict sites.

Further **secondary** research will be available from your local Planning Office. This may help you with your research on the **existing site** such as its former use, and date of dereliction (this would be useful in an ecological survey to assess the speed of colonisation). The Planning Office can also help with information on **future plans** for the use of various sites and **reclamation strategies**.

Scale of visual pollution | Table 3.23

Aspects of pollution	No pollution 0	1	Badly polluted 2	3	Guidelines
Size of site			✓		The larger the size, the worse the pollution.
How obvious to passers-by			✓		The more obvious the pollution, the higher the score.
Colour and texture		✓			Heaps of rubbish are worse than a thin cover.
Impact on surrounding area		✓			If in a pleasant area the pollution scores higher than if in a poor area.
State of buildings	✓				Old, derelict buildings nearby score highly.
Abandoned cars, etc.	✓				Many large items such as old beds, fridges, etc. score highly.
Litter			✓		The greater the density, the higher the score.
Smells	✓				The worse the smell, the higher the score.
Vegetation cover			✓		Plants can hide pollution, so the greater the vegetation, the lower the score.

This has been completed for a derelict land photograph. A score of 0 means the item is not present.

WEBSITES

Air quality information and statistics for the UK is available at: www.aeat.co.uk/netoen/airqual

The BBC website includes information on pollen: www.bbc.co.uk/weather

Ceefax and Teletext also have air quality data.

TRANSPORT INVESTIGATIONS

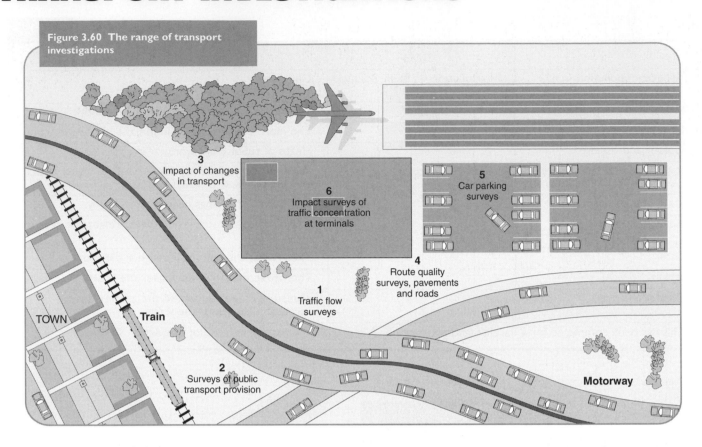

Figure 3.60 The range of transport investigations

3 Impact of changes in transport

6 Impact surveys of traffic concentration at terminals

5 Car parking surveys

4 Route quality surveys, pavements and roads

1 Traffic flow surveys

TOWN Train

2 Surveys of public transport provision

Motorway

Transport networks play a fundamental role in modern life – they get you to work, to school, to shop and on holiday. If things go wrong and the networks do not work, this causes major problems for a government. There are currently major concerns about congestion and the environmental problems caused by motor cars especially in urban areas, and how to develop quality public transport systems such as a well run, safe national railway system. Any large scale developments such as a new airport terminal or motorway cost billions of pounds and stir up an enormous amount of protest from local people. Wherever you live, you are bound to find many interesting and relevant investigations on transport. Figure 3.60 gives you some ideas – your problem therefore is finding the most appropriate focus.

TRAFFIC FLOW SURVEYS

Basic surveys

The basic field survey for a traffic flow project involves measuring the volume of passing traffic from a roadside vantage point, to focus on daily, weekly or possibly seasonal variations in traffic flow.

Where you do your traffic surveys will depend on your title focus. For an individual project a focus on a difficult road junction or a major or complex roundabout is ideal, but even then you really need a group of friends to cope with the volume of rush hour traffic and its movements. Alternatively for a traffic survey in a small town you will need to record traffic flows at different points at the same time so you need a group (Figure 3.61).

You will need to devise a suitable booking sheet (Table 3.24), recording the time period, vehicle classes, and any other supplementary information you require. Make sure that all members of the group

TOP TIPS

▶ Get the scale correct – if you are working as an individual you will not be able to cover traffic flow issues in a large town. Even for a small town you will need a group of friends to do the basic traffic surveys to cover all roads **simultaneously**.

▶ Do the necessary background reading. The Highways and Transport Departments in your chosen area have all sorts of policy statements and surveys which you can consult to help you interpret your primary surveys.

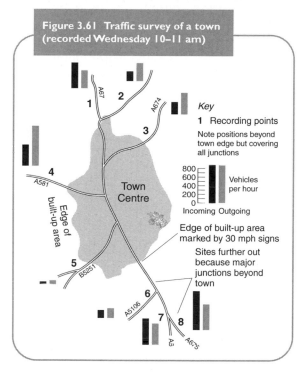

Figure 3.61 Traffic survey of a town (recorded Wednesday 10–11 am)

Key

1 Recording points

Note positions beyond town edge but covering all junctions

	Vehicles per hour
800 600 400 200 0	

Incoming Outgoing

Edge of built-up area marked by 30 mph signs

Sites further out because major junctions beyond town

You should then decide what supplementary surveys you might need to do. These could include the following.

▶ The **speed** of selected traffic. Time a sample of vehicles between a visible distance, e.g. between two lamp posts (usually about 200 metres apart). Convert the timing in seconds to kilometres per hour.

▶ The **homebase** or key description of distinctive vehicles can be noted and the time recorded at which they passed the survey point. These vehicles will then serve as **indicator vehicles** when you are trying to assess speed of flow through a congested town centre.

▶ **Occupancy rates** can be useful to assess the potential for car pooling in cutting congestion. Record the number of occupants per car over a period of time, e.g. 35 minutes during rush hour.

▶ **Noise levels** can be recorded using a decibel recorder (noise meter) as part of an environmental impact project.

collect the information in the same way. Be very clear about the directions of flow and use a mechanical counter for areas of very busy fast moving flows.

Traffic census booking sheet **Table 3.24**

Name of recorder J. Core		Time from 10.45am to 11.00 am		Day Wednesday		Date Oct. 21st
Position of site Garstang Rd (at Princess Rd turn off)		Town Preston		Weather Fine, sunny		

Class 1 Bicycles		Class 2 Motorbikes, scooters		Class 3 Cars (incl. three wheelers)		Class 4 Light vans, minibuses		Class 5 Heavy vehicles, lorries, tractors, buses, etc.		Sources of lorries, bus numbers, etc.
In	Out	In	Out	In	Out	In	Out	In	Out	
II	III	II	IIII	ЖЖЖ ЖЖ III	ЖЖЖ ЖЖ ЖЖ III	ЖЖЖ I	ЖЖ II	ЖЖЖЖ II	ЖЖ III	Garstang, Bolton, Kendal, Accrington, Nelson

- Traffic flows can be converted to an **index of volume** per minute where each vehicle type is worth a different score: bikes = 1 point, motorbikes = 2 points, cars = 3 points, vans/minibuses = 4 points, lorries/buses = 6 points.

- The **saturation** levels of roads. Use either an index of volume or an estimate of passenger car units (PCUs), e.g. a lorry/coach is said to be worth 3 PCUs, a light van 2 PCUs and a car 1 PCU. You can then use Table 3.25 to assess whether the road has reached saturation point. If parking is allowed on the road this will mean the road is narrower so saturation levels will be reached more easily.

- **Fuel consumption survey**s can also be used when trying to evaluate public transport alternatives to private cars. You will need to note the car type and specification to get an indicator of its fuel consumption. The figures in Table 3.26 are based on an average speed of 15kph – typical of heavy urban traffic.

The latest Department of Environment scales for vehicle tax introduced in 2001 could also be useful.

The section on Pollution – pages 180–191 also shows you how to look at the impact of traffic.

Figure 3.62 shows how you can use annotated graphs to present the results of your surveys.

Additional surveys

Once you have undertaken your basic traffic surveys you could aim to answer one of the following questions.

- What factors influence any differences of flow at a particular point (daily, weekly, etc.)?

- What factors lead to congestion at a particular point?

- Why are traffic flows a problem at a particular point?

- What impact does the weather have on traffic flow at a particular point?

- Has the new layout at a particular point been successful?

- What impact does street parking make on traffic flow?

- Has a particular road reached saturation point (or is there a need for a bypass)?

- What environmental problems might result from the heavy traffic flows on a particular road?

You could also undertake a **traffic trouble spot** survey. Trouble spots are areas where vehicles are held up for at least 1 minute, either stationary, or crawling. Difficult junctions, roundabouts, traffic lights and road works are obvious causes, but often surprise factors such as burst water mains, or factories closing down for their summer holiday can also contribute.

Saturation points of different road types — Table 3.25

Road Class	Complete width in metres	Saturation point in PCUs per hour, measured both ways
Single lane highway (with passing places)	4	<200
2 lane country road B road	7.3	375
Three lane A road (central overtaking)	10	700
Dual carriageway	14	1500
6 lane	22	3000
8 lane motorway	30	4000

Fuel consumption and CO_2 emissions of different vehicles — Table 3.26

Vehicle type	Fuel consumption (litres per kilometre)	Approximate CO_2 emission (grams per km)
Motorbike	0.08	200
1000cc car	0.09	220
2000cc car	0.11	250
Van/minibus	0.12	300
Bus	0.40	1000
HGV	0.53	1400

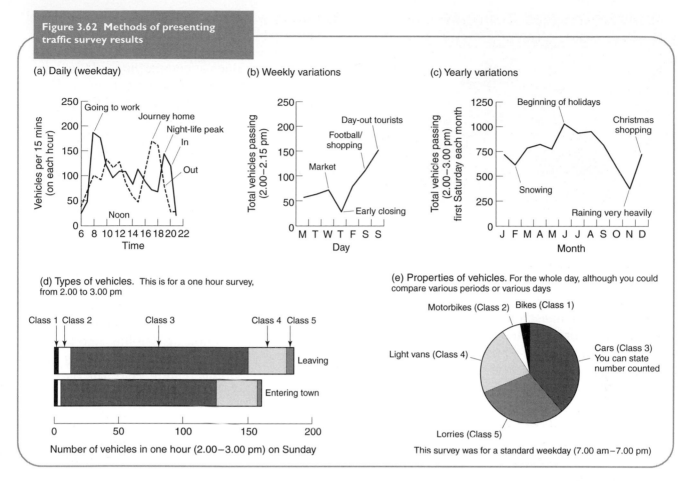

Figure 3.62 Methods of presenting traffic survey results

(a) Daily (weekday)

(b) Weekly variations

(c) Yearly variations

(d) Types of vehicles. This is for a one hour survey, from 2.00 to 3.00 pm

(e) Properties of vehicles. For the whole day, although you could compare various periods or various days

In order to carry out a trouble spot survey you will need to do the following.

- Walk through the town at rush hour, mapping the delay sites.

- Carry out synchronised surveys at all delay points, timing average delays, i.e. waiting times and length of tailbacks, and recording traffic flows and pedestrian flows where relevant.

- Draw large diagrams to account for the delays at each point. Suggest possible traffic management solutions.

- Support your work with a **perception survey** of congestion points or on traffic delays in getting to work.

You could take your project further and look at the issues involved in traffic management.

1. **Accident (black) spots** Figures can be obtained from the Highways and Transport Departments and police departments of accidents in your

chosen town over a period of time. Once you have done a map of the **distribution of road** accidents (a dot map would be ideal) you can analyse them by time, type, day, week, season and weather to pick out accident spots where there are multiple incidents. Visit the site, record the basic traffic flow, congestion and speed and then if possible interview local residents as to the possible causes. Carry out your own risk assessments of the sites to assess the potential hazards and dangers. Research the local press for causes of any accidents and record the weather conditions at the time of accidents.

At present excessive speeding on **rural roads** is a major issue. You could carry out speed surveys and traffic flow/congestion surveys of all the roads in a designated rural area. A **road density index** may produce some interesting data, especially when related to number of cars in one area.

2. **Traffic calming measures** Such measures are designed to slow down traffic and reduce the risk

Figure 3.63 Traffic calming methods

of accidents, e.g. by speed cameras, speed restriction signs, bumps, rumble strips, coloured roads (Figure 3.63). You can take a particular area (village or town) and assess the success of various schemes. **Speed** measurements will provide the main data here **or** you could pose the question whether certain problem areas need a speed camera or traffic calming. You can also survey the residents of your chosen area as to their opinions about the issue.

3. **Journey generator issues** Areas such as school gates, retail parks, office/factory entrances or sports events can cause major hazard and congestion zones at certain times of the day, especially where no off street parking or drop off and turning zones are provided. Make maps of pedestrian flows, traffic flows, parking patterns and bus surveys for the zone in question.

   You could support your investigation with questionnaire surveys such as that shown in Table 4.27.

4. **Road safety and efficiency within a residential area** Inner city areas and council estates are often home to high densities of population and contain many old aged pensioners and large numbers of young children. Both groups are vulnerable to traffic accidents. Street quality profiles in such areas will reveal major shortcomings in safety due to parked cars inhibiting visibility, and large numbers of children playing in the street because of a lack of gardens (Figure 3.64). The parked cars add to congestion (you could calculate

A possible questionnaire to support a journey generator investigation	Table 4.27

Which of the following factors most strongly influence your decision to travel to work by car?

Tick the boxes that apply:

a) A convenient and secure parking space is generally available ☐

b) I enjoy driving and usually encounter little congestion or delay on the journey to work ☐

c) The car is essential most days for the type of work I do ☐

d) I have a car sharing arrangement with work colleagues which reduces the costs ☐

e) Public transport is not a viable option in terms of:

   1 Public transport stops are not close to my home/workplace ☐

   2 Unsuitable frequencies/timetables ☐

   3 Cost ☐

   4 Personal dislike ☐

   5 Crowded (may have to stand) ☐

f) I need the car to drop/collect children at school and/or for shopping/visits before/after work ☐

g) I enjoy the freedom a car gives me to listen to the radio/music, or to just think and relax ☐

h) I am not particularly concerned about my own contribution to air pollution ☐

i) I feel safer travelling in my own car ☐

j) Any other reason .... ☐

k) I don't travel to work by car because .... ☐

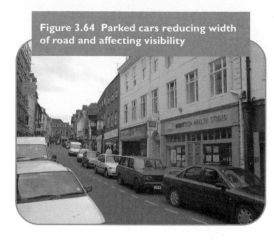

Figure 3.64 Parked cars reducing width of road and affecting visibility

congestion indexes with and without the parked cars) as many roads, especially in inner city areas, are used as journey to work runs by people trying to make short cuts to avoid congestion points on the main roads. A combination of traffic flows, speed measurements, congestion measurements, parking surveys, street quality surveys, accident rate surveys, people activity surveys, and residents' questionnaires provides a superb range of data to highlight the problem. Solutions by the council may well be under consideration, so you could also evaluate these, or alternatively devise your own plans for traffic calming and management to improve the area.

PUBLIC TRANSPORT SURVEYS

Since 1986 there have been significant changes in public transport provision such as the deregulation of bus routes and the privatisation of the rail network. In some places these changes are for the better, but in some areas they make things worse.

The following is a list of possible issues you could investigate.

▶ Problems of rural bus provision (dispersed, low customer threshold, high costs). Look at the impact of deregulation and the recent grants for rural buses such as Sunday services. Also investigate the role of the village minibus and shared taxis.

▶ Improving estate bus provision in an urban area (look at two contrasting estates).

▶ Solving the problems of bus congestion in a large city – in particular reviewing the success of bus lanes.

▶ The role of park and ride schemes – evaluating their success and potential.

▶ The development of an integrated bus/rail system in part of a large urban area– evaluating its success.

▶ The creation of a new rapid transit or super tram system.

▶ The opening of a new commuter station, or improved service on a train commuter line.

▶ The creation of green transport policies such as the role of cycle ways.

You could develop a series of specific questions to layer your research.

▶ How and why does bus or train frequency vary in an area?

▶ How does bus frequency relate to settlement size and length of journey?

▶ How does bus frequency between settlements relate to the theoretical interaction statistics?

▶ How does tourism impact on bus or train frequency and use?

For each of these investigations bus or rail timetables will provide the raw data. Figure 3.65 shows you how to develop topological route maps and flow maps to show the daily/weekly **frequencies** of buses. You can compare the level of **service and frequency** on a number of routes and then calculate the bus coverage index as shown below.

$$\text{Index of bus coverage between towns A and B} = \frac{\text{Journey time from A to B (minutes)}}{\text{Number of buses between A + B in both directions}}$$

The lower the index value, the greater the density of buses along the route.

Bus usage surveys can either be done as **loading** surveys at the terminus (particularly useful for park and ride schemes where you can also monitor the service car park) or by making a journey. You will need to get prior permission as bus and rail stations are private land – often the transport companies have their own surveys which they will make available to you once you contact them. Try to record the age, gender and passenger profile (shopping bags are a give away) at various times of the day on a number of routes.

Figure 3.65 a) Rose diagram showing the main towns served by the Skipton bus service. b) Flow map showing the number of buses coming into Skipton

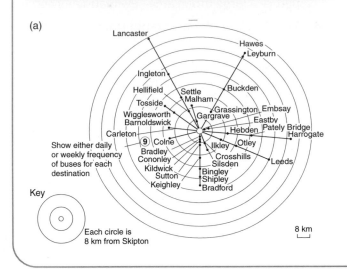

(a)

Key

Each circle is 8 km from Skipton

8 km

Show either daily or weekly frequency of buses for each destination

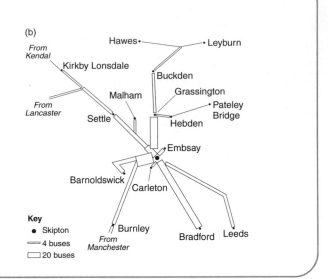

(b)

Key
• Skipton
⇒ 4 buses
▭ 20 buses

Travelling the routes gives you a clear picture of occupancy rates, passenger profile and turnover (Figure 3.66). Research secondary data such as the fare structure (relate ticket changes to distances travelled) as well as any usage figures. The local council will provide information on subsidies and many also have results of user surveys. You can also conduct a **passenger** questionnaire on usage, safety and quality of service.

Detour indexes can be used to compare journey directness and time, compared to car times as shown below.

$$\text{Detour index} = \frac{\text{Journey time/distance by bus}}{\text{Journey time/distance by car}}$$

The lower the index the better the bus route (Figure 3.67).

Connectivity indexes may be very useful when analysing the provision of rural or suburban bus networks to assess their relative efficiency. With numerous small private operators and a lack of integrated transport this may have become worse since deregulation. Figure 3.68 shows you how to work out the connectivity index. The range of values is from 0.5 (poor connectivity) to 3 (excellent connectivity).

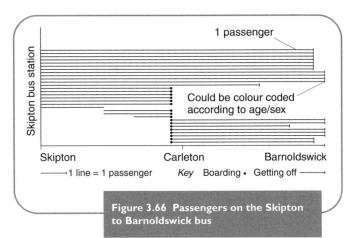

Skipton bus station

1 passenger

Could be colour coded according to age/sex

Skipton   Carleton   Barnoldswick

1 line = 1 passenger   Key   Boarding • Getting off

Figure 3.66 Passengers on the Skipton to Barnoldswick bus

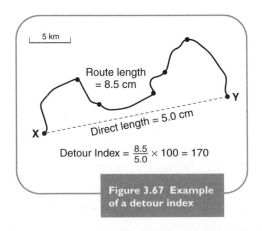

5 km

Route length = 8.5 cm

Direct length = 5.0 cm

X   Y

Detour Index = $\frac{8.5}{5.0} \times 100 = 170$

Figure 3.67 Example of a detour index

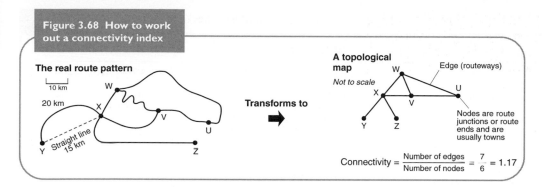

Figure 3.68 How to work out a connectivity index

The real route pattern

A topological map
Not to scale

Transforms to

Edge (routeways)

Nodes are route junctions or route ends and are usually towns

Connectivity = $\dfrac{\text{Number of edges}}{\text{Number of nodes}} = \dfrac{7}{6} = 1.17$

Changes in public transport networks

Projects considering change in a transport network require the use of historic timetables and other supporting documentation such as old maps, so **check** their availability for your chosen area before you start. The same basic surveys as above can be done at appropriate date intervals (bus frequency, coverage, connectivity, accessibility, etc.).

In some areas of Britain the network has changed dramatically, for example in the 1960s and 70s many branch railway lines were **closed** (Beeching closures). In one or two cases the lines have been reopened, usually as a restored railway tourist attraction. More usually the railway tracks became disused, and now form sites of ecological value with footpaths or cycle tracks.

Surveys to assess the ecological value of a disused railway line include quadrat surveys and height of vegetation surveys. To assess the potential of a disused railway line, view assessment surveys and footpath quality surveys can be carried out. Questionnaires could also be devised to gather information and opinions from users of the railway line or people living near by.

New transport developments such as a new super tram route or the building of a by-pass can be investigated at proposal stage, at building and development stage, and at operational stage, where the basic surveys will need to be combined with an environmental impact assessment survey. For such a project you should carry out a visual impact survey by walking along the proposed routes to assess the quality of the landscape. Use computer graphics to construct images which show the likely environmental impact once the road is built. Use maps to record the quality of features and buildings, footpaths and roads which will be lost. Conduct noise surveys (from an existing similar route) to work out likely noise impacts. Use a matrix to work out the likely improvements of journey times the new road will bring (Table 3.28).

Matrix to show the impact of M65 motorway on car journey times, between selected destinations

Table 3.28

Time before / Time after M65	Selected places close to M65				
	Barrowford	Bury	Fence	Manchester	Rawtenstall
Brierfield	0.21 / 0.10	0.52 / 0.56	0.05 / 0.05	0.76 / 0.71	N/A / N/A
Burnley	0.30 / 0.18	0.47 / 0.45	0.13 / 0.13	0.72 / 0.70	0.36 / 0.18
Colne	0.25 / 0.32	0.71 / 0.52	0.23 / 0.23	0.96 / 0.77	0.58 / 0.36
Nelson	0.12 / N/A	0.61 / 0.49	0.06 / 0.13	0.86 / 0.74	0.48 / 0.26
Padiham	0.25 / 0.15	0.51 / 0.39	0.08 / 0.08	0.76 / 0.62	0.28 / 0.11

(Left axis label: Selected places on M65)

1. Measure shortest distances between two centres before and after motorway.
2. Calculate times using D. of E. figures in either minutes or parts of an hour. Assume no traffic jams.
3. Assess results in terms of losses and gains in journey times.
4. Motorway times may not be shorter because of journeys to interchanges.

The costs and benefits of a new bypass	Table 3.29

COSTS	BENEFITS
Trade will drop in town centre businesses. **Evidence:** Car parking surveys, pedestrian surveys, business questionnaires, e.g of shops and garages.	Will take away traffic flow from busy town centre and make it more pleasant. **Evidence:** Traffic census, before and after noise surveys.
Environmental damage to the countryside. **Evidence:** Pollution, noise surveys, traffic flows; questionnaires.	Improved journey times. **Evidence:** Questionnaires and matrixes.
Walkers' routes are blocked by it. **Evidence:** Footpath surveys. Wildlife carnage in the countryside area. **Evidence:** Surveys of dead animals on road.	Reduce number of accidents. **Evidence:** Accident figures.

For the **construction** phase concentrate on environmental impact surveys of dust, noise, lorries and visual disamenity.

If the road has been built an ideal framework is to look at costs and benefits. Table 3.29 shows how to tackle this for a new bypass.

You can write up the analysis under environmental and socio-economic costs and benefits. In most cases the overall impacts will be good after the by-pass has been built, but there may be major issues for shops which lose their passing trade because of the old road losing so much traffic.

ROUTE QUALITY SURVEYS

Route quality surveys can be carried out on pavements, roads, public footpaths and cycle ways. The method is the same for all. You should walk round your chosen area and decide on the need for a sampling strategy or not. At the same time prepare a booking sheet to evaluate **quality**. Carry out detailed quality surveys and use photographs to calibrate your scale. Support this with appropriate in-depth interviews of residents, users, and also perception surveys to highlight problems. Activity surveys (paths), pedestrian surveys (pavements) and traffic surveys (roads) can also be undertaken to indicate usage.

For **road surveys** you should record the following information.

- Surface type (noise level).

- Incidence of bumps, cracks and holes.

- Occurrence of hostile litter, e.g. glass/nails.

- Quality of drains and kerbs.

- Traffic volumes, speed, noise and dust surveys.

For **pavement surveys** you should record the following information.

- Surface type – condition/flatness.

- Kerb height and condition.

- Quality of street lighting.

- Incidence of animal waste and litter.

- Occurrence of weeds.

- Degree of landscaping.

- Obstacles for blind or disabled users, e.g. wheelie bins and street furniture.

- Guidance for blind users, e.g studs.

Paths or cycle way surveys can be assessed in terms of the following.

- Signposting and parking access.

- Aesthetic quality – plant and animal species diversity.

- Underfoot conditions.

- Degree of maintenance, e.g. undergrowth, stiles, gates, foot bridges, obstacles and hazards.

- Links to other footpaths, service provision.

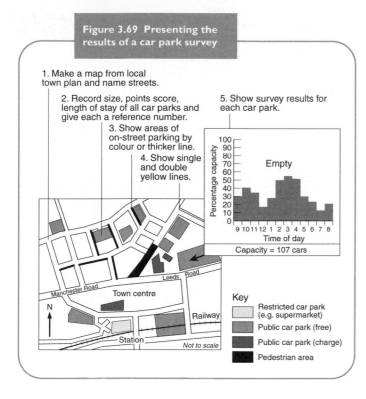

Figure 3.69 Presenting the results of a car park survey

1. Make a map from local town plan and name streets.

2. Record size, points score, length of stay of all car parks and give each a reference number.

3. Show areas of on-street parking by colour or thicker line.

4. Show single and double yellow lines.

5. Show survey results for each car park.

Empty

Capacity = 107 cars

Town centre

Manchester Road

Leeds Road

Railway

Station

Not to scale

N

Key

Restricted car park (e.g. supermarket)

Public car park (free)

Public car park (charge)

Pedestrian area

PARKING SURVEYS

Car parking surveys are very straightforward. Set yourself the task of assessing the provision, usage and adequacy of car parks and on street parking within a chosen small town.

As can be seen from Figure 3.69 the first stage is to undertake an **initial survey** to record areas of parking, restricted parking and no parking. Map the number of spaces, meters and areas of double and single yellow lines. At each car park record the prices, length of stay allowed and total number of spaces. For a large town a group will be needed to do this.

You will need to devise a scale for assessing the standard of each car parking area. The following are likely to be relevant:

▶ Ease of access.

▶ Ease of layout (i.e. getting into the space).

▶ Cost.

▶ Safety and security for both driver and car (look for security lighting, cameras etc.).

▶ Size.

▶ Likelihood of spaces (look for waiting cars).

▶ Overall convenience of access to town centre.

▶ Quality of surface.

▶ Amenities such as toilets, information, litter bins, lifts.

▶ Access for disabled.

▶ General ambience and environmental quality.

You could devise a scale of 0–5 points to score the criteria, and then total them to give an overall quality index; the higher the figure the better the provision. You may decide to **weight** certain factors more than others.

You will need to carry out **user** surveys at various times of the day and days of the week to assess occupancy rates. Illegal parking may indicate saturation point has been reached in prime locations, so record any examples. Also try to carry out a sample duration survey in short stay areas, logging the registrations of cars at 30 minute intervals.

When writing up describe the provision with maps (Figure 3.69) and discuss the strategy behind the pricing (usually high pricing for short stay car parks close to the CBD, and low cost long stay parks on the edge of the town centre for workers). Secondly describe and explain your usage surveys and then discuss how adequate the supply of car parks appear to be.

INVESTIGATING THE IMPACT OF TRANSPORT TERMINALS

A project looking at the impact of ports, airports or large stations can be especially interesting if there has been a recent change. **However**, it is not always easy to get access to all the usage data (for security reasons), and there are few opportunities for you to carry out your own surveys in restricted areas.

If you undertake **an airport impact study** your basic information will usually consist of an **airport handbook** (with **some** useful data) and **flight schedules**. It is also useful to arrange an interview with the airport manager in order to gain information on passenger and freight movement daily, monthly and annually. From this you can construct destination maps (use flow lines). As the schedules give types of aircraft you can work out the seating capacity for each route. You can also graph all the movements daily, weekly, seasonally and annually. If a new terminal has been built or a runway extended you can assess the impact by comparing aircraft data and destinations before and after.

Try to obtain information on flight paths for landing and take off so you can carry out noise surveys at various times of the day. Dust and pollution surveys may also be possible – check with the Environmental Health Department for any existing data. You can then map severely affected housing areas. The council will have maps of those households eligible for grants towards triple and double glazing to avoid the noise.

In terms of measuring airport capacity, an activity survey in the airport (with permission), car park, and traffic flow survey on the approach roads will be useful at various times. It is also possible to record airport **capacity**, i.e. how many planes are parked at various times of the day, using large scale maps.

Questionnaire surveys of flight path residents and other people in the area will give you information on the environmental impact, usage and employment at the airport.

WEBSITE

www.detr.gov.uk

DETR website hold data on regional development.

SPORT, LEISURE, RECREATION AND TOURISM ECOSYSTEMS

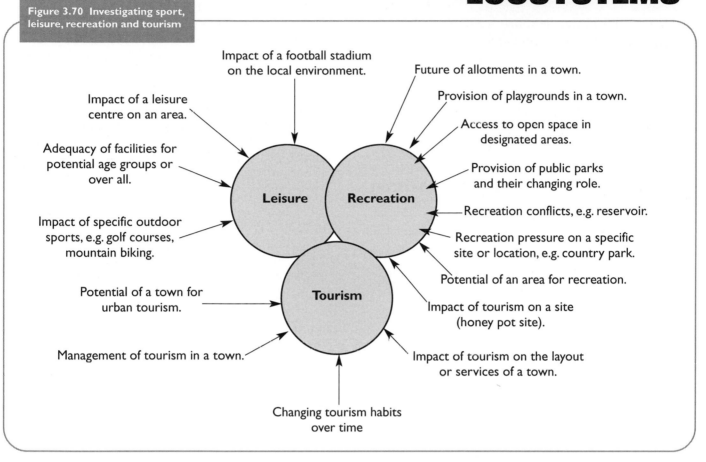

Impact of a football stadium on the local environment.

Future of allotments in a town.

Impact of a leisure centre on an area.

Provision of playgrounds in a town.

Adequacy of facilities for potential age groups or over all.

Access to open space in designated areas.

Leisure

Recreation

Provision of public parks and their changing role.

Impact of specific outdoor sports, e.g. golf courses, mountain biking.

Recreation conflicts, e.g. reservoir.

Recreation pressure on a specific site or location, e.g. country park.

Potential of an area for recreation.

Potential of a town for urban tourism.

Tourism

Impact of tourism on a site (honey pot site).

Management of tourism in a town.

Impact of tourism on the layout or services of a town.

Changing tourism habits over time

As you can see from **Figure 3.70** there is an enormous range of possible investigations which fall under the heading sport, leisure, recreation and tourism.

Leisure includes all non-work related activities; **recreation** is a pursuit engaged in during leisure/spare time; **sport** implies an organised, active leisure pursuit; **tourism** is all activity involving an overnight stay, whilst most is for recreation, some tourism is for business purposes. If you decide to undertake a project in this area you will need to focus in on a particular aspect of sport, leisure, recreation or tourism.

Look in the local papers for a key issue such the impact of motorbike use on a local common, or the plans to move the 'home' of your local football club to a new location. When looking at tourism go for a precise manageable focus. For example, the impact of tourism in Scarborough could be converted to the impact of **beach tourism** on the environment or economy, or shops and services of Scarborough.

BASIC TECHNIQUES

Most enquiries in this field involve looking at the supply of facilities, the demand for them, and the impact that the use of particular facilities in particular locations makes on an area. In a rapidly changing industry, changing locations of supply, and changing demands of consumers are also worthwhile areas of investigation.

Assessment of the supply of provision

This kind of study can focus on the quantity, spatial distribution, and **quality** of the provision of facilities.

Facilities surveys

The **quantity** of facilities in an area can initially be assessed using by using the Yellow Pages and other directories and maps to identify and locate all the facilities within your proposed study area.

The next stage would be to visit all your possible sites to confirm what is actually there as directories and maps are frequently out of date.

For services such as parks and swimming pools you can usually obtain a leaflet which gives you details about the facilities, costs of use etc. and you can supplement this by making a detailed map of each. Use similar scales for all facilities surveyed and annotate your maps in as much detail as possible.

It is vital at this stage that you decide on the limits of your survey area. This will depend on the scale of your project. For **high order** facilities such as leisure centres, swimming pools or cinemas you will need to survey a large town, urban borough or a whole rural district. Organising travel around the area is of prime importance. For more **localised** facilities such as parks, allotments or playing fields a medium sized town of 20,000 – 100,000 would be ideal.

For facilities such as open spaces (both formal and informal), the usage will be very localised so one or two wards would be suitable for detailed analysis (Figure 3.71).

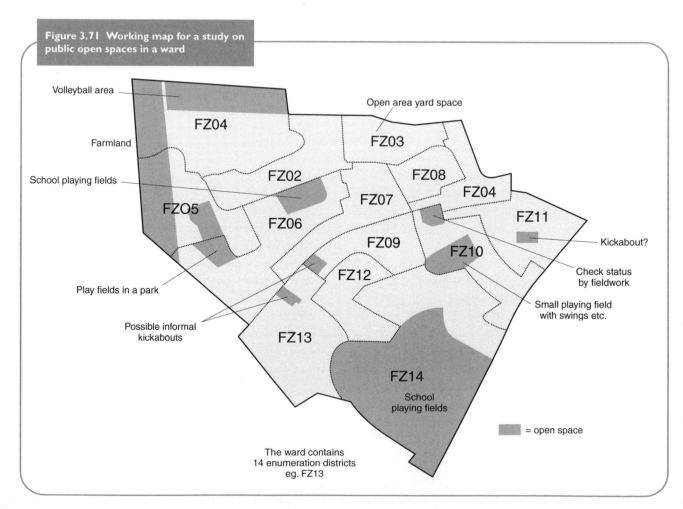

Figure 3.71 Working map for a study on public open spaces in a ward

| | Booking sheet for an open space survey | | | | | | | | | | | | | | | | | | | | | | | | | | Table 3.30 |

✓ Facility available — Park	Size in hectares	Children's	Football pitch	Rugby pitch	Cricket pitch	All weather area	Bowling green	Tennis courts	Grass play area	Hockey pitch	Pitch and putt	Athletics track	Picnic site	Boating lake	Paddling pool	River frontage	Ornamental gardens	Café/kiosk	Sports pavilion	Bandstand	Toilets	Wood	Casual landscape	Others	Centrality index	Rank	Hierarchy level
1. Towneley	150	2	1 / 3		1 / 1		3	3	2				✓			✓	✓	✓	✓		✓	✓	✓	Stately Home Golf C			
2. Queens	20						2	6	✓		✓						✓						✓				

You will need to prepare **booking sheets** to record what is present (Table 3.30). This information can then be used for developing a hierarchy of provision.

Quality surveys

During your survey you should make an assessment of the **quality** of provision. Table 3.31 shows you how to do this for a swimming pool; the criteria you use can obtained by questioning 10 regular swimming pool users as to what factors influenced their choice of a location for a swim.

The advantage of a quality survey is that the results can all be obtained by just observation; the disadvantage is that it is based only on opinion.

Table 3.32 shows a useful quality survey for the condition of play spaces and playgrounds.

User surveys

User surveys form a very important part of both facilities and quality surveys. There are a number of ways you can undertake these surveys.

▶ Obtain (where available) statistics for hourly, daily, weekly and seasonal usage of the facilities. Most sports centres keep these records as part of their accounts.

▶ You can supplement these secondary statistics by undertaking your own primary intensity of use survey. **Footfall** surveys at the entrance can be combined with **visual** surveys at regular intervals for all the facilities (Figure 3.72). A **capacity** survey can be carried out to assess the percentage fullness – for example a limit is set for safety reasons on swimming pool use, and some facilities like tennis and bowls have a physical playing limit (Figure 3.73).

▶ Enquire about membership lists and club usage. From the membership list you can construct a catchment area for the facility.

	Devising a quality survey for a swimming pool		Table 3.31
1. Interview 10 regular swimmers to establish what factors they think are important when choosing a pool to swim in.	2. List the criteria ▶ Size of pool ▶ Cleanliness of water ▶ Quality of changing area ▶ Range of facilities ▶ Beginners' pool ▶ Safety features ▶ Temperature of water ▶ Ease of access by bus ▶ Car parking ▶ Opening times ▶ Variety of swim sessions ▶ Cost per swim ▶ Age of pool	3. Scale of criteria ▶ Use a scale of 0–5. ▶ Weight the important criteria × 2. ▶ Decide on maximum number of points that can be awarded.	4. Devise the booking sheet.

Place	Structural damage	Paint peeling	Graffiti	Rotting timber	Parts not working	Total
The Brampton	8	8	8	8	8	40
Guernsey Drive	8	8	8	8	8	40
Westbury Park	5	5	5	0	5	20
Cotswolds Ave	5	8	5	8	8	34
Acacia Ave	2	2	2	2	2	10

Quality of playgrounds — Table 3.32

Maximum total posible = 40

Guide

8 = None – the playground is in immaculate condition.

5 = Little – there are small amounts of damage but they are not noticeable/do not make the playground unsightly or unsafe to use.

2 = Some – there is some damage but it does not affect equipment all over/prevent it from working, yet if there was slightly more damage the playground would be unsafe.

0 = Much – the playground has damage all over it and is severely deteriorating and would be classified as unsafe for children.

A superb technique to undertake is **an activity survey**. This would be most useful for a large, open air facility such as a reservoir, country park, beach or tourism honey pot site. Figure 3.74 shows you how to record individuals at a particular time by age, sex and activity. You could analyse the distribution **spatially** using nearest neighbour analysis to assess the degree of clustering and assess the impact of **distance decay** from a key facility such as a visitor centre car park. It is also possible to divide the survey area into grid squares and then correlate the density of users (using Spearman's Rank correlation coefficient) with **landscape quality** or beach quality results and the **recreational potential index** readings (see below).

In order to carry out a **recreational potential survey** use an OS map and evaluate the land in each grid square as follows:
  – List all the services which might attract tourists (accommodation, shops, etc.).
  – List all the recreational amenities (picnic sites, footpaths, fishing rivers, etc.).
  – List all the scenic and historic attractions (woods, historic houses, view points, etc.). Rate each of the above on a scale from 1–5 in terms of how important you consider them **for tourists**. A caravan site or information centre would score 5, but a post office or public seat might only score 1.
  – To each score add an Index of Accessibility, scoring 5 on an A road, 4 for a B road, and 3 for an unclassified road passing through each kilometre square.

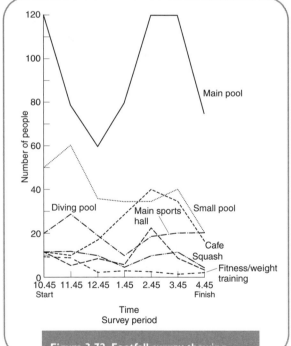

Figure 3.72 Footfall survey showing intensity of use of facility

Figure 3.73 Capacity survey showing usage of tennis courts

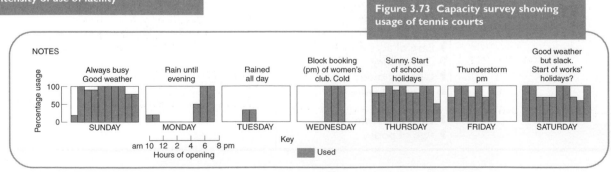

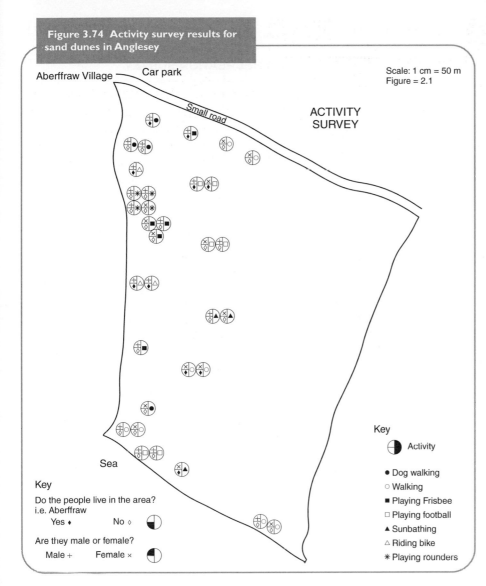

Figure 3.74 Activity survey results for sand dunes in Anglesey

Aberffraw Village — Car park

Small road

Scale: 1 cm = 50 m
Figure = 2.1

ACTIVITY
SURVEY

Sea

Key

Activity

● Dog walking
○ Walking
■ Playing Frisbee
□ Playing football
▲ Sunbathing
△ Riding bike
✳ Playing rounders

Key

Do the people live in the area?
i.e. Aberffraw

Yes ◆   No ◇

Are they male or female?

Male +   Female ×

score for that square to see to what extent visitors use the areas with high potential.

Questionnaires

Questionnaires can be extremely useful for the following surveys investigating the supply of provision.

▶ **Catchment survey**: You will need to interview up to 100 people and ask them where they have come from.

▶ **Attraction survey**: This is a more in depth survey where you should ask up to 25 people what they are planning to do or have done at a particular site, and what are their opinions about the quality of a particular facility (you could use **your** quality survey here).

You could even set your questionnaire out with a map and actually ask people to draw in the route they took, and what attractions they stopped at (Figure 3.76).

▶ **Tourists v residents questionnaire**: This can be very useful when assessing the **impact** of a particular development or the importance of various management issues such as car parking in a honey pot village.

— Certain elements of the landscape detract from its tourist potential (pylons, eyesores, military camps), although some both attract and repel tourists (nuclear power stations). For each ugly element, subtract up to 3 points from the square total score.

▶ Make a final evaluation of the land by allocating it a score in one of the following classes: high, medium, low and unsuitable tourist potential. Draw a map showing the grid squares and shade each according to its potential (Figure 3.75). The next step is to undertake a field survey of the number of tourists using each square, and for what reason they are using it. Correlate the number of visitors with the overall

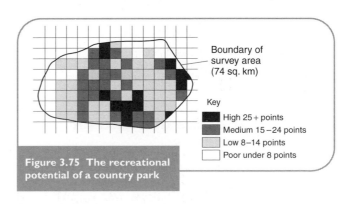

Boundary of survey area (74 sq. km)

Key

■ High 25 + points
▨ Medium 15–24 points
▧ Low 8–14 points
□ Poor under 8 points

Figure 3.75 The recreational potential of a country park

Figure 3.76 Questionnaire map

We shall be most grateful if you will assis our project by completing this questionnaire. The answers are for statistical puropses only. You are not required to give your name.

KEY TO MAP

- • Town/village
- ⬛ Reservoir
- ━ Major road
- ─ Minor road

WRITE here your approximate time of arrival on Dartmoor

| a.m. |
| p.m. |

QUESTIONS TO BE ANSWERED ON THE MAP:–

1) MARK with arrows your complete route thus ──→ ──→

2) MARK your stopping places along the route (with an X) and by each mark put an appropriate letter and number(s):–

3) MARK the number of people in your party/family group

Age	Nos.
0–14 years	
15–24 years	
25–64 years	
65 or over	

LETTER – HOW LONG YOU SPENT

A	up to ½ hr.
B	about ½–1 hr.
C	about 1–3 hrs.
D	over 3 hrs.

NUMBER – WHAT YOU DID

1	rested only
2	stopped for picnic/refreshments outside
3	stopped in cafe/pub/tea shop
4	admired view/photography
5	swim/sunbathe/water activity
6	stopped for children to play
7	looked at plants/wildlife
8	other (state here)

4) Where have you come from today? _____

5) Where is your usual place of residence, if different from 4) above?

_____

If you have any particular comments you wish to make on DARTMOOR or this questionnaire, please do so overleaf. THANK YOU VERY MUCH FOR YOUR CO-OPERATION

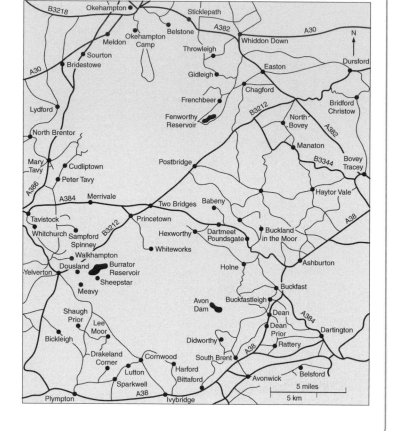

DARTMOOR RECREATION SURVEY

TOP TIPS

Be prepared to develop a range of **primary** field work methods. Many investigations rely far too much on poorly designed questionnaires which are only given to a small sample of people. Questionnaires will be necessary but the questions must be carefully targeted directly to your chosen enquiry.

Assessment of the demand for particular facilities

An assessment of demand is a common choice of A level project title. The adequacy of services needs to be defined in **absolute** terms, i.e. meeting national standards and also in **relative** terms, e.g. by looking at the percentage of open spaces by ward. It can also be defined in user terms (**capacity** surveys) and by questionnaire to assess opinions on quantity and quality of facilities. Adequacy is often variable by age or even gender and therefore appropriately designed and sampled questionnaires are a vital component in such studies .

One important type of survey is an **access** survey. You can either use random number tables to generate sampling points within a designated area, or use specific sampling points such as villages and hamlets in a rural area, or particular streets in an urban area to develop **access wheels** (Figure 3.77). this will help you find out how far (distance or time) and in what direction people have to travel for a range of basic leisure services (playing field, swimming pool, sports centre, cinema etc.). You can do the surveys by age (youth, middle aged, pensioner) and record total distance and average distance travelled for services.

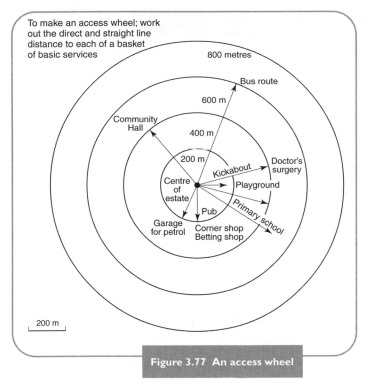

To make an access wheel; work out the direct and straight line distance to each of a basket of basic services

800 metres

Bus route

600 m

Community Hall

400 m

200 m

Kickabout

Doctor's surgery

Centre of estate

Playground

Pub

Primary school

Garage for petrol

Corner shop
Betting shop

200 m

Figure 3.77 An access wheel

Assessment of the impact of a proposed/existing development on the surrounding environment

There are a number of options available under this heading, again many involve observational work as well as questionnaires.

Visual impact surveys

These surveys can be particularly useful for a proposed project such as a new football stadium. Digital camera images can be used (or artists' drawings) to actually quantify the change of long range view brought about by the new development (Figure 3.78). Alternatively you can use a landscape/visual quality index.

Environmental impact surveys

These are now a compulsory part of any planning proposal, so people have the opportunity to assess the likely effects. To undertake an impact survey of your own collect the baseline data by visiting the site to identify the likely environmental factors which could be affected. Read all about the proposal and try to identify any changes that might happen. It is always worth visiting a similar development to try to quantify likely impacts. Draw up a **pilot**

Figure 3.78 The view before and after development

View 2 – before

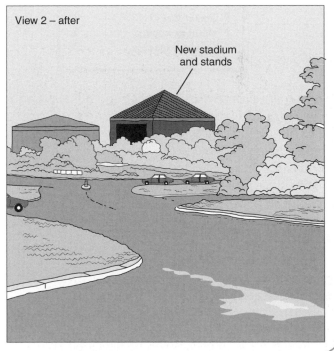

View 2 – after

New stadium and stands

	Pilot environment impact assessment								Table 3.33
	Types of changes which could happen with the project								**Total**
Characteristics of the environment which might be affected	Habitat modification	Urbanisation	Waste problems	Loss of farmland	Localised traffic	Noise	Decreasing water quality		
Modification of habitat (flora & fauna)									
Impact on hydrology (drainage)									
Impact on landscape/visual quality									
Impact on pollution, either air or water									
Impact of noise and vibration									
Impact on traffic flow (roads/footpath)									
Impact on employment opportunities									
Impact on recreational opportunities									
Impact on land prices									
Impact on existing land use		2							
Total									I

Insert single score
▸ **No impact**   0
▸ Low impact   I
▸ Medium impact   2
▸ Severe impact   3

environmental impact assessment and fill in the grid (Table 3.33). If you are very lucky and the project goes ahead during the year you are doing your enquiry you may be able to undertake one assessment for the construction phase and one for the completed project. If the project/development remains at the planning stage you will have to visit a similar site with a similar development and use a variety of surveys to **model** the potential impact.

Negative externalities

You should carry out a questionnaire survey to develop nuisance **fields** in which you can map the intensity of the nuisance posed by identified problems such as a football ground (Figure 3.79). An overall nuisance contour map can then be developed by either totalling the nuisances or averaging the results.

The results of the questionnaires are plotted for each nuisance. Using an outline map draw an 'isoline' map for values 1 and 2. An isoline joins points of equal value, e.g. contours.

You will need to 'rough-out' these lines before you attempt the finished shape. Watch out for islands of values and do not cross other isolines.

In your analysis

1. Describe the pattern of 'nuisance fields' for the different problems.

2. Attempt an explanation of why the effect of the stadium is uneven.

Figure 3.79 Nuisance fields questionnaire

Good afternoon/morning; we are from _____ school/college and we are undertaking a survey of this area as part of our 'A' level <u>geography</u> course. Would you mind answering one or two questions?

Living here, do you find, on average, that when _____ Football Club are playing at home (name ground if necessary) they create:

(Tick box)   No Nuisance   A Nuisance   A Serious Nuisance

            ☐ score 0   ☐ score 1   ☐ score 2

2. Would you say that any of the following are "nuisances" or "serious nuisances" (score 0 for "no nuisance", score 1 for "nuisance", 2 for "serious nuisance")

    TRAFFIC _____

    PARKED CARS _____

    PEDESTRIANS _____

    HOOLIGANISM/

    VANDALISM _____

    NOISE _____

    OTHER (please specify)

    _____

Thank you for your help

EXAMPLE PROJECTS

Impact of a football stadium

Context

Most large stadiums were built around 50 years ago near city centres in the heart of residential areas. There are many negative externalities (such as noise, traffic, car parking) on match days when crowds of over 30,000 are involved. Many clubs are considering a move to a new out of town purpose built location. You can explore the need for this, and also in some cases the impact of the change on the old site, or the success or otherwise of the new site.

Secondary data

Obtain statistics for past seasons for all home matches and identify two contrasting matches to survey from the current season's programme, e.g. cup match, weekday evening match.

Set up interviews with a member of the club's administration, the secretary of the supporters club, and local police unit (to assess policing issue).

Primary surveys

- Undertake car parking and traffic flow surveys before, during and after match.

- Use a decibel recorder to measure noise at selected distances away from ground.

- Carry out housing condition, environment quality and house price surveys in the area immediately a round the stadium.

- Carry out a visual impact survey of the ground from a variety of angles and distances.

- Carry out pedestrian flow surveys just before a match on routes leading to the ground.

▶ Record vandalism/graffiti on the main routes followed by supporters.

▶ Carry out an impact of flood lights survey for an evening match.

▶ Interview a member of the club's administration, the secretary of the supporters' club and the local police unit (to assess policing issues).

▶ Interview the owners of local stores and garages to carry out **footfall** surveys on match and non match days for these services.

▶ Interview the community development officer at the club to record the club's involvement in local schemes.

▶ Obtain employment figures for local people at the club.

Always do control surveys on a non match day so you can assess any impact against them.

Key questions

▶ How and why does the football club affect the physical environment?

▶ What do the local residents feel about the stadium?

▶ What is the extent of the effect of the football club on the local economy?

▶ Is the football club a benefit or a burden to the area?

▶ Is the current location suitable for a large stadium or is an out of town location more suitable?

Golf course development

Context
All over the world there is an enormous demand for new golf courses and many hundreds are constructed

every year. There are numerous issues associated with the type of land the course is built on – there may be loss of ecological, agricultural or historical land. The impact of the golf course on the surrounding area is also important – golf courses generate a lot of increased traffic. There are also implications for water use as watering the greens requires an enormous quantity of water.

Figure 3.80 shows the general planning guidelines for the construction of a new golf course.

Fieldwork

▶ Assess the need for a golf course. Devise surveys to locate the existing courses and to look at their usage (activity surveys, catchment surveys). Question golfers as to the need for a new facility and the quality of existing facilities.

▶ Assess the **proposal** and quantify the loss of land. You will need to undertake field mapping to look at land capability, hedgerow loss, heritage loss, tree loss, etc. to quantify the costs. Visits to existing golf courses can be made to assess the benefits (landscape quality survey, employment etc.). Questionnaire surveys on local residents to assess their views on the issue can be conducted.

▶ An impact analysis will need to be carried out using an **environmental impact assessment** (Table 3.33). Visits to existing similar sized working golf courses can be made to look at the likely impact of increased traffic flows.

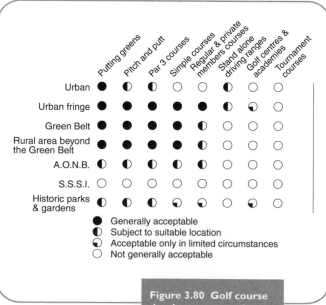

Figure 3.80 Golf course development matrix

Key issues

The key issues revolve around a benefit-cost analysis related to the specific site. Often the sites chosen are greenfield attractive sites as brownfield sites have too many problems.

Impact of sports/leisure centre on the surrounding area

Context

Sport and leisure centres are designed to benefit the local community – they are sometimes attached to a local high school for this reason. Often the local residents do not make use of the centre and they complain about its impact. Building a new sports centre causes controversy as to whether it is really needed and why it should be built on a particular site.

Field work

▶ For a proposed centre you should assess the landscape quality of the site before and after it is built. For an existing centre, map the centre and carry out **activity and capacity surveys**.

▶ A residents' questionnaire (using stratified sampling) will need to be undertaken. Look at the impact of traffic flows, noise, litter, parking etc. at various times of the day. If it is a new centre find an existing one to model the likely numbers that will use the centre.

▶ If you are investigating an existing centre obtain details of visitor usage, catchment area and carry out a quality assessment so you can see how it fits in to the local sports provision hierarchy.

Key issues

The key questions will be: do the benefits outweigh the cost, and are the local residents or users satisfied? A questionnaire on possible improvements is a good idea. Figure 3.81 shows the results of one such survey.

Surveys of parks and open spaces

Context

Parks are just one facet of a whole range of open spaces from playing fields to outdoor running tracks.

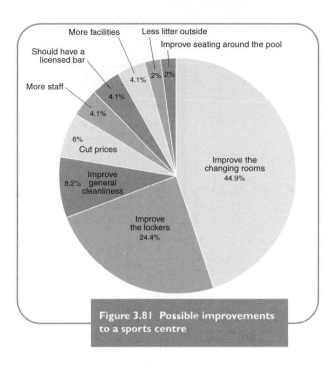

Figure 3.81 Possible improvements to a sports centre

They were the pride of Victorian England with wonderful floral displays and bandstands. Today, because of escalating maintenance costs some have become run down places of fear where glue sniffers and other undesirables congregate. The type of entertainment they provide is less popular (although pram walking, jogging and strolling are still very common) as the provision of cars has enabled people to go to National and Country Parks.

Often there is inequality in the distribution of open spaces with many poor areas being especially deprived of quality open spaces.

Methodology

▶ Map and classify all the open spaces – Table 3.34 shows a hierarchy based on park size and range of facilities.

▶ Find the centrality index for each open space.

$$\text{Centrality value of a facility} = \frac{\text{number of each facility in all parks}}{100}$$

To find the centrality index for a park add up all the centrality values for the specific facilities (Figure 3.82).

▶ Conduct landscape quality surveys and activity surveys.

A hierarchy of parks and open spaces   **Table 3.34**

		Size	Facilities	Catchment
Level 1	Neighbourhood park	1–2 ha.	Kickabout and playground	Radius $\frac{1}{2}$ km
Level 2	Local park	Up to 20 ha.	Basic football pitch, bowls, tennis	Radius 1–2 km
Level 3	Large park	Up to 50 ha.	Full range of sports facilities	Radius 4–9 km
Level 4	Main park	Usually 100 ha. +	Full range of sports, bandstands, gardens, museums	City-wide and some tourist potential from outside area

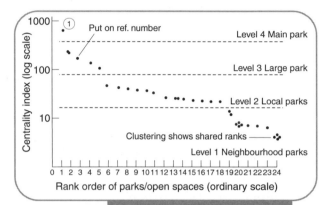

Figure 3.82 Centrality index of a town's open spaces in relation to their rank order

▶ Conduct user surveys to find catchment areas, and residential questionnaires to find out local people's reactions to their nearby open spaces.

▶ Carry out environmental surveys of graffiti, vandalism and litter.

▶ Assess the distribution of parks, open spaces or play grounds by ward (Table 3.35) and in relation to national levels of provision.

Possible hypotheses

▶ Parks and public open spaces are distributed evenly throughout the city.

▶ There is a wide range of public open spaces in terms of size and amenities showing a clear hierarchy.

▶ Conduct quality surveys of all the facilities and also capacity surveys of features such as putting and bowls.

The percentage of parks and public open spaces in the wards of Carlisle   **Table 3.35**

WARD	No. of squares for ward	Area (m²)	No. of squares for parks	Area (m²)	% Area of park in ward
BELAH	82	3,820,000	17.5	700,000	21.34%
BELLE VUE	43	1,720,000	4.5	180,000	10.47%
BOTCHERBY	57	2,280,000	15.5	620,000	27.19%
CURROCK	26.5	1,060,000	1	40,000	3.77%
DENTON HOME	60	2,400,000	7.5	300,000	12.50%
HARRABY	54	2,600,000	1	40,000	1.85%
MORTON	31.5	2,160,000	6.5	260,000	20.63%
ST. AIDAN'S	27.5	1,100,000	3.5	140,000	12.73%
STANWIX URBAN	92	3,680,000	20	800,000	21.74%
TRINITY	49	1,960,000	11	440,000	22.45%
UPPERBY	33.5	1,340,000	6	240,000	17.9%
YEWDALE	42.5	1,700,000	3	120,000	7.06%

- The parks are used by people of different ages and gender and for different reasons.

- General public opinion about parks is complimentary, but recently some parks have become zones of fear.

- The parks are experiencing major management problems.

- There is a relationship between the range of amenities and the quality of parks and visitor numbers.

Recreational usage and potential conflicts at a particular site (e.g. reservoir, beach, national park honey pot)

Context

With growing numbers of people seeking recreation, where a facility can be used for a variety of purposes. Inevitably some conflicts will occur especially in peak periods. For example, noisy power boats impact on nesting birds. In some cases the conflicts are extremely dangerous, for example when jet skiers can threaten the lives of swimmers. The conflict is frequently resolved by user codes and zoning, which of course has to be policed. You could investigate what the conflicts are, or how they could be managed. A good investigation is to actually review the efficacy of a management plan.

Methodology

- Having decided on your location, the first stage is to carry out a reconnaissance survey and visit the information/visitor centre to make an appointment with a warden or development manager. She/he may have secondary data such as a management plan and detailed user figures as well as regulations/zoning strategies.

- You will need to carry out a number of **activity** surveys at various times of the day and in various weather conditions. Classify the recreational activities into active/passive,

formal/informal and low/medium/ high impact, e.g. walking a dog on the beach would be active, informal and medium impact.

- Make an impact matrix to show how the various activities conflict with the environment.

- Make a conflict matrix as shown in Table 3.36.

- Carry out basic questionnaires to find out the duration of stay of visitors, distance travelled to site, frequency of visit, reason/motive for visit, opinions about the site (litter, noise, erosion, overcrowding, vandalism etc.). From these basic questionnaires you can map comparative spheres of influence etc.

Impact of tourism at a honey pot site

Context

Many popular sites at the coast or in National Parks are visited in huge numbers with a wide range of consequent problems, especially on peak summer Sundays and Bank Holidays (Figure 3.83). This over-use has an environmental impact, and also a socio-economic impact on the people living in the area – most of these do get an income from tourism, but dread the busy summer season. There are also some long term concerns such as the rising costs of housing, and the skewed provision of services. Key issues include why your chosen area has become a honey pot and also what impact the growth of tourism has made on the environment and local community.

		Conflict matrix				Table 3.36		
✓ = No conflict ? = Possible conflict X = Conflict		Walkers	Dog walkers	Cyclists, roller-bladers etc.	Anglers	Rowers	Sunbathers	Picnickers
Walkers		–	?	X	?	✓	?	?
Dog walkers		?	–	X	X	?	X	X
Cyclists, rollerbladers etc.		X	?	–				
Anglers					–			
Rowers						–		
Sunbathers							–	
Picnickers								–

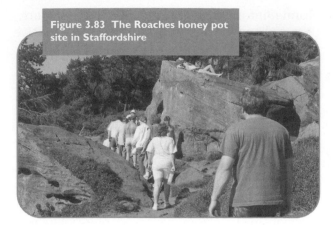

Figure 3.83 The Roaches honey pot site in Staffordshire

▶ Make sure you take relevant photographs so you can fully annotate them.
▶ Make sure your mapping of footpath erosion, parking overflow and visitor numbers is very detailed.
▶ Present your results in interesting ways (Figure 3.84).

Methodology

Honey pot surveys are enhanced by having a control, i.e. a non honey pot day such as a wet Monday in Autumn with which to make comparisons.

▶ Map the site, carrying out visual quality and landscape quality surveys in each gridded square. Use a land use classification scheme to record all services etc.

▶ Carry out **activity** surveys, car parking surveys (see Fig 3.84) and traffic surveys at various times of the day.

▶ Carry out footpath erosion and usage surveys as well as pedestrian counts in a village.

▶ At the end of the day carry out a litter survey. Note any instances of vandalism.

▶ Interview tourists (aim for a sample of 100) on the purpose and duration of their stay as well as key management issues. Also interview up to 50 residents to find out about local employment in tourism and the perceived problems.

▶ Contact the appropriate agency to see whether there are any management plans in place, for example for traffic and car parking.

high value urban areas. Some allotments appear untended and overgrown and in some authorities there are more allotments than would-be cultivators. To some people, however, they represent a vital income generator, a chance to grow organic home grown produce, or a wonderful leisure time activity. The value of this is difficult to quantify.

Methodology

▶ Find out where the allotments are located in your chosen town – the council will have a list of owners for each site.

▶ Visit the sites and carry out landscape evaluation and development potential surveys. This will enable you to see whether the allotments contribute to the quality of the environment. It could be followed up by questionnaires asking residents backing on to the sites to see how they feel about the use.

▶ Carry out sample land use surveys of individual allotments to try to assess if there would be any contribution to individuals' vegetable supplies. Assess the quality of cultivation (use photographs such as Figure 3.85) and devise a descriptive scale.

Allotment surveys

Context

Allotments are under threat from land developers as they are a low value land use in

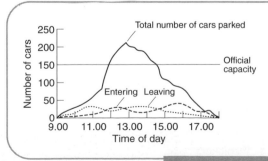

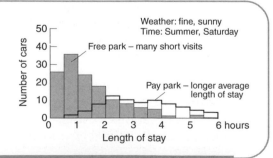

Figure 3.84 Results of car park surveys in a National Park

Figure 3.85 The quality of cultivation in an allotment

▶ Carry out a questionnaire on allotment owners (see Fig 3.86). Find out details/characteristics of the owners, the nature of cultivation and their reasons for cultivating. Try to interview a 10 per cent sample from each set of allotments.

Key issues
You should analyse the distribution, assess the costs and benefits and evaluate the future role of allotments.

Site:

Plot No.

HISTORY OF ALLOTMENT USE

For how long have you had this allotment?

Have you ever had an allotment elsewhere? Where (Cheltenham/other town)? When (years)?

CHARACTERISTICS OF THE OWNER

What is your nationality/the area of Britain from where you originate? How long have you lived in this area?

Do you have a garden at home? If yes what are its dimensions?

What is/was your occupation, profession or trade?

Do you work full-time or part-time, or are you unemployed or retired?

Are you an employee or self employed?

Estimation of owner's age

0–20   21–40   41–60   60–80   over 80

CULTIVATION OF THE ALLOTMENT

How often do you visit your allotment and in which season?

What crops did you grow in 1994?

Did you grow any new crop in 1994 for the first time?

Approximately how much do you spend on your allotment per year? ....

Is this expenditure covered by the value of the crops you produce?   Yes/No

ENVIRONMENTAL ISSUES Have you gone organic?

                                        In the past

Do you use (a) Horse or other natural manure?
            (b) Chemical fertiliser?
            (c) Neither?
            (d) Both?

                                        In the past

Do you use (a) fungicides?
            (b) insecticides?
            (c) pesticides?

Figure 3.86 Allotment owner questionnaire

Urban tourism

Context

A number of opportunities exist for a study on urban tourism – if Stoke on Trent and Bradford can develop successful tourism so can many other towns! Assess the potential or review the management of existing tourism developments. Urban tourism has been a very significant way for towns to earn income and provide jobs because of deindustralisation in the manufacturing sector.

Methodology

▶ Obtain leaflets advertising the attractions (churches, shops, museums, view points). You may find that the council have developed a tourism plan as part of their economic development plan. Also you may be able to obtain a useful accommodation list and some tourist statistics and surveys from the council.

▶ Map the attractions and devise quality surveys for them. Visitor books, e.g. from churches, are very useful when working out where tourists have come from. Map the distribution of pubs, cafés, restaurants, hotels and other special tourist facilities (e.g. buses, stations, car parks etc.). Collect details of possible tours so you can map the frequency of visits to attractions.

▶ Carry out a detailed visitor questionnaires such as the one shown in Figure 3.87.

▶ Use the tourist leaflets to see how easily you can find the attractions, whether the literature is accurate and informative, and if the attractions are value for money. You can also assess the quality of any guide services etc.

▶ Assess the effectiveness of the management plan for urban tourism in your town if one exists, and then make up your own modified management plan in the light of your field observations.

The impact of tourism on the layout of facilities in a town

Context

Many towns which specialise in tourist activity have a completely different zoning structure to other towns – especially if they are lakeside or seaside towns. The zoning is based on the cost of the land, with large hotels and facilities competing for sea front space. Tourism-orientated towns tend to have higher numbers of shops and services per inhabitant than other towns.

Methodology

▶ Make a detailed land use map identifying all the tourist facilities.

▶ If you are looking at the impact of tourism on services and shops you can grade the shops and services as: of no value to tourists (1), partially orientated towards tourists (2), only selling tourist goods (3).

▶ You can also carry out pedestrian counts, traffic flows, and car parking surveys during the tourist season and compare these with similar surveys conducted out of season.

▶ Interview residents, tourists and shopkeepers to assess the impact of tourism on shops and services.

Figure 3.87 North Shropshire visitor questionnaire

1. Have you ever visited North Shropshire before?
If Yes, was it for a day visit or overnight?

No: Never been before ☐   Yes: Day Visit ☐   Overnight ☐

2. How many times have you visited North Shropshire in the past?

Once ☐   Twice ☐   More ☐

3. Why did you decide to visit North Shropshire this time?

Recommendation	☐ 1	Advertisement	☐ 4
Press/PR/Radio/Articles	☐ 2	Been Before	☐ 5
Magazine/Newspaper feature	☐ 3	Other	☐ 6

_____

4. What was your **main** reason for visiting North Shropshire this time?

Visiting friends/family	☐ 1	Sightseeing	☐ 5
Attending an event/show	☐ 2	Shopping	☐ 6
Commercial/Business trip	☐ 3	Visiting an attraction	☐ 7
Activity:(delete as appropriate)	☐ 4	Location	☐ 8
Sport Walking Cycling		Other	☐ 9
Riding Fishing Other		_____	

5. What type of information did you obtain on North Shropshire before or during your visit?

Essential Shropshire	☐ 1	
Shropshire Tourism Magazine	☐ 2	
Holiday Brochure	☐ 3	Name _____
General Guide Book	☐ 4	Name _____
Web information	☐ 5	
Other	☐ 6	Name _____
None	☐ 7	

6. Where did you obtain this information from?

Shropshire Tourist Information Centre	☐ 1
Other Tourist Information Centre	☐ 2
North Shropshire Tourism	☐ 3
Shropshire Tourism	☐ 4
Shropshire Tourism website	☐ 5
Other website	☐ 6
Village Shop/Church/Pub	☐ 7
Where you were/are staying	☐ 8
Other	☐ 9

_____

_____

7. Length and date of stay in North Shropshire this time?

Date of visit _____
Day Visit ☐ 1   Overnight ☐ 2   Number of nights? ☐

8. Where did you/are you staying this time?

Bed & Breakfast	☐ 1	Guest House	☐ 5
Self Catering	☐ 2	Boat/Marina	☐ 6
Hotel	☐ 3	Friends	☐ 7
Caravan or Camping	☐ 4		

9. What is the address of your accommodation?

_____

_____

10. Please grade your accommodation accordingly

Excellent ☐   Very Good ☐   Good ☐   Fair ☐   Poor ☐
Any other comments _____

11. Was the advertising description of your accommodation accurate?

Yes ☐ 1   If No ☐ 2   Please state why _____

12. Was your accommodation fairly priced?

Yes ☐ 1   No ☐ 2

13. Will you revisit North Shropshire?

Yes ☐ 1   No ☐ 2   Why? _____

14. If you have seen a copy of Essential Shropshire, how would you rate it?

Good ☐ 1   Average ☐ 2   Poor ☐ 3
Comments _____

_____

15. What did you use Essential Shropshire for?

Accommodation ☐ 1   General Information ☐ 3
Places to visit ☐ 2   Other _____

16. From the list below please tell us:

	If you have visited any of the towns listed	If yes, how did you rate them (score 1 low to 5 high)
ELLESMERE	Yes/No	
MARKET DRAYTON	Yes/No	
WEM	Yes/No	
WHITCHURCH	Yes/No	

17. What improvements/further facilities would you like to see? (Specify particular town or wider area)

_____

_____

INDEX

Access wheels 209
Acid rain 185
Activity surveys 207
Anemometer 89
Aspect 50, 74
Association index 120

Beach temperature profiles 101
Beach transect 63
Beaufort Scale 89
Biological indicators 43
Biotic Index 44–5
Break point 166–7
Building age survey 143, 156
Building decay index 157
Building functions 143
Building height survey 151
Bulk density (soil) 110
Burglarability index 176
Bus index 197
Bypass surveys 200

Cailleux's Index 79–80
CBD structure 148
Census 162, 169
Centrality index 213–14
Cinema index 13
Cliff profile 57
Clinometer 32
Cloud symbols 90
Cloud types 90
Coastal protection schemes 66
Connectivity index 199
Cost-benefit analysis 48, 66, 200
Crime statistics 177
Cross-sections 51

Delimiting the CBD 149
Deprivation index 162
Detour index 198
Discharge calculation 33
Diversity index 114, 150

Ecological impact index 123
Electoral register 173
Elongation ratio 75
Environmental impact assessment 210
Erosion index 58

Facility survey 204
Field sketches 72, 78
Footfall surveys 153, 154
Friction table 31

Golf development matrix 212
Gradients 49
Graffiti assessment 178

Hinterland surveys 166
House price survey 162
Housing condition 157
Housing decay index 11
Housing density 157
Hydraulic radius 32
Hypotheses 39

Impact matrix 69, 128, 199, 215
Infiltration rates 54
Interception 54, 99

Journey generators 196

Kick sampling 45

Land capability index 14
Land use (and key) 134–5, 155
LEAP reports 42
Lichen chart 185
Litter maps 189

Meander maps 40
Mental maps 16
Mobility maps 175

Nearest neighbour analysis 17, 151, 173
Nitrates 46
Nitrogen dioxide 185
Nodality index 142
Noise scale 187

Open space survey 205
Oral histories 15
Orientation 81
Oxygen levels 46
Ozone (low level) 185

Pantometer method 51
Parking restriction survey 153
Pedestrian surveys 151, 206
Perception surveys 177, 195
pH 46, 103, 107
Population pyramid 143
Power's scale 79–80
Precipitation measurement 88
Provenance studies (ice) 73, 77

Quadrats 22, 112
Quality indexes 13, 14, 205, 206
Quality of life survey 160

Questionnaires 9, 67, 122, 136, 144, 145, 183, 196, 208, 211, 217, 219

Recreational potential 206
Relative humidity 89
Rhan's Index 83
River corridor symbols 48
River velocity 29
Road saturation levels 194
Rocks (recognition) 58
Roundness index 80
Runoff maps 54

Sample size 7, 9, 79, 114
School transport survey 172
Sheltergrams 95
Shopping hierarchy 154
Shopping survey 152
Shoreline Management Plans 68
Sinuosity index 40
Site evaluation table 141
Smoke Index (Ringelmann) 184
Soil colour chart 107
Soil erosion map 111
Soil profile 103
Soil texture graph 110
Spearman's rank 38, 75, 117, 163, 167
Standard error 9
Stemflow 100
Stone board 80
Storm simulation 54
Stream competence 34
Street survey index 12
Succession 115
Suspended sediment 35

Traffic counts 192–4
Traffic trouble spots 194
Trampling index 127
Transects 94, 133, 158
Transmission coefficient 98
Turbidity 46

Urban heat island 96

Visibility map 90
Visual pollution index 191
Voting patterns 178

Wave energy calculation 60
Weathering 83
Wetted perimeter 30
Windspeed/direction 89

Zingg shape 81
Zonation 115